ATTRACTING PhDs TO K-12 EDUCATION

A DEMONSTRATION PROGRAM
FOR SCIENCE, MATHEMATICS,
AND TECHNOLOGY

Committee on Attracting Science and Mathematics
PhDs to K-12 Education:
From Analysis to Implementation

Center for Education
Division of Behavioral and Social Sciences and Education

and

Division of Policy and Global Affairs

National Research Council

NATIONAL ACADEMY PRESS
Washington, DC

NATIONAL ACADEMY PRESS • 2101 Constitution Avenue, N.W. • Washington, D.C. 20418

NOTICE: The project that is the subject of this report was approved by the Governing Board of the National Research Council, whose members are drawn from the councils of the National Academy of Sciences, the National Academy of Engineering, and the Institute of Medicine. The members of the committee responsible for the report were chosen for their special competences and with regard for appropriate balance.

This study was supported by Grant No. 1000316 from the Burroughs Wellcome Fund, Grant No. 99-4245 from the William and Flora Hewlett Foundation, Grant No. D00001 from the Carnegie Corporation of New York, and Grant No. SG-99-134 from the Camille and Henry Dreyfus Foundation to the National Academy of Sciences. Any opinions, findings, conclusions, or recommendations expressed in this publication are those of the author(s) and do not necessarily reflect the views of the organizations or agencies that provided support for the project.

International Standard Book Number 0-309-08427-X

Additional copies of this report are available from National Academy Press, 2101 Constitution Avenue, N.W., Lockbox 285, Washington, D.C. 20055; (800) 624-6242 or (202) 334-3313 (in the Washington metropolitan area); Internet, http://www.nap.edu

Suggested citation: National Research Council. (2002). *Attracting PhDs to K-12 Education: A Demonstration Program for Science, Mathematics, and Technology.* Committee on Attracting Science and Mathematics PhDs to K-12 Education. Center for Education, Division of Behavioral and Social Sciences and Education and Division on Global and Policy Affairs. Washington, DC: National Academy Press.

Printed in the United States of America

Copyright 2002 by the National Academy of Sciences. All rights reserved.

THE NATIONAL ACADEMIES

National Academy of Sciences
National Academy of Engineering
Institute of Medicine
National Research Council

The **National Academy of Sciences** is a private, nonprofit, self-perpetuating society of distinguished scholars engaged in scientific and engineering research, dedicated to the furtherance of science and technology and to their use for the general welfare. Upon the authority of the charter granted to it by the Congress in 1863, the Academy has a mandate that requires it to advise the federal government on scientific and technical matters. Dr. Bruce M. Alberts is president of the National Academy of Sciences.

The **National Academy of Engineering** was established in 1964, under the charter of the National Academy of Sciences, as a parallel organization of outstanding engineers. It is autonomous in its administration and in the selection of its members, sharing with the National Academy of Sciences the responsibility for advising the federal government. The National Academy of Engineering also sponsors engineering programs aimed at meeting national needs, encourages education and research, and recognizes the superior achievements of engineers. Dr. Wm. A. Wulf is president of the National Academy of Engineering.

The **Institute of Medicine** was established in 1970 by the National Academy of Sciences to secure the services of eminent members of appropriate professions in the examination of policy matters pertaining to the health of the public. The Institute acts under the responsibility given to the National Academy of Sciences by its congressional charter to be an adviser to the federal government and, upon its own initiative, to identify issues of medical care, research, and education. Dr. Harvey V. Fineberg is president of the Institute of Medicine.

The **National Research Council** was organized by the National Academy of Sciences in 1916 to associate the broad community of science and technology with the Academy's purposes of furthering knowledge and advising the federal government. Functioning in accordance with general policies determined by the Academy, the Council has become the principal operating agency of both the National Academy of Sciences and the National Academy of Engineering in providing services to the government, the public, and the scientific and engineering communities. The Council is administered jointly by both Academies and the Institute of Medicine. Dr. Bruce M. Alberts and Dr. Wm. A. Wulf are chairman and vice chairman, respectively, of the National Research Council.

COMMITTEE ON ATTRACTING SCIENCE AND MATHEMATICS PHDS TO K-12 EDUCATION: FROM ANALYSIS TO IMPLEMENTATION

M. PATRICIA MORSE, *Chair*, University of Washington, Seattle
MARGARET COZZENS, University of Colorado at Denver
ARTHUR EISENKRAFT, Bedford Public Schools, Bedford, NY
DANINE EZELL, San Diego City Schools
EMILY FEISTRITZER, Center for Education Information, Washington, DC
MARIA LOPEZ FREEMAN, California Science Project, Monterey Park, CA
MYLES GORDON, American Museum of Natural History, New York, NY
VICKI JACOBS, Harvard University, Cambridge, MA
DAVID A. KENNEDY, Office of the Superintendent of Public Instruction, Washington (retired)
MARY LONG, University of Texas, Austin
JOHN A. MOORE, University of California at Riverside (Emeritus)*
N. RONALD MORRIS, University of Medicine and Dentistry of New Jersey
MAYNARD V. OLSON, University of Washington, Seattle (until February, 2001)
KRISTINA PETERSON, Lakeside School, Seattle, WA
ERIC ROBINSON, Ithaca College, Ithaca, NY
JAMES H. STITH, American Institute of Physics, College Park, MD
KIMBERLY TANNER, University of California, San Francisco

JAY B. LABOV, *Deputy Director*, Center for Education
KEVIN D. AYLESWORTH, *Study Director*, Center for Education
GEORGE REINHART, *Senior Program Officer*, Division of Policy and Global Affairs
TERRY HOLMER, *Senior Project Assistant*, Center for Education

---

*Deceased May 2002.

Dedication
John A. Moore
(1915–2002)

John A. Moore was Professor of Biology Emeritus at the University of California, Riverside and a member of the National Academy of Sciences. His long and active scientific career spanned seven decades, beginning when he published his first research paper as a teenager and continuing until his death. His last book, *From Genesis to Genetics*, was published in 2001. He is perhaps best known for his unending devotion to science education, which he served at all levels from classroom instruction to textbook writing (including the highly regarded Biological Sciences Curriculum Study texts for high school—still in print after 40+ years—and the *Science as a Way of Knowing* series for university-level instruction). He also served the National Academies and other professional and educational organizations on committees and study groups too numerous to detail. Dr. Moore was especially devoted to defending the teaching of evolution as an essential component of any complete modern curriculum. His ideas and high standards had a major influence on the deliberations of this committee and the ideas advanced in this report. He will be sorely missed.

Contents

PREFACE		ix
EXECUTIVE SUMMARY		1
1	INTRODUCTION	5
	A Need and an Opportunity, 5	
	The Committee's Task, 7	
2	THE CHALLENGES OF EDUCATION	10
	The Need for High-Quality Teachers, 10	
	New Approaches for Professional Development, 12	
	Beyond Professional Development, 15	
3	FROM GRAD SCHOOL TO GRADE SCHOOL	17
	Are They Interested?, 17	
	What Do They Know?, 20	
	What Do They Need to Learn?, 21	
	How Can They Learn What They Need to Know?, 23	
	How Can They Be Retained in K-12 Education?, 24	

| 4 | PROPOSED DEMONSTRATION PROGRAM | 26 |

    A National 2-Year Fellowship, 27
        A National Program, 27
        Program Duration, 30
        Teacher Preparation and Certification, 30
        Support for the Fellows, 32
    Other Program Characteristics, 33
        Recruitment, 34
        Selection and Placement, 35
        Mentoring and Leadership Preparation, 37
    Structure, 39
    Funding, 40
    Evaluation, 41
    Next Steps, 42

| REFERENCES | 43 |

APPENDICES

A	Executive Summary: Attracting Science and Mathematics PhDs to Secondary School Education	51
B	Agenda and Participants: Workshop on Attracting PhDs in Science and Mathematics to Careers in K-12 Education	63
C	Nontraditional K-12 Teacher Preparation Programs	72
D	Programs to Strengthen Connections Between Professions and K-12 Education	82
E	Gallup Teacher Perceiver Instrument	87
F	National Residency Matching Program	89
G	Biographical Information on Committee Members	91

Preface

A significant number of new PhDs in science, mathematics, and engineering are seeking careers that make good use of their hard-earned skills, but are outside of the traditional ones at universities or in industry. Some of them wish nothing more than to teach, to pass on the passion of their research, to share their connections to the wonders of discovery, and to use their talents to bring an understanding of the promises and limitations of science, mathematics, and engineering to others. That desire sometimes turns to schools—from kindergarten through high school (K-12)—perhaps because their parents or siblings are teachers or because they realize that university life or life in industry does not appeal to them. The special talents and learning they have accumulated in their doctoral studies make them a potentially marvelous resource for the nation's K-12 system of education.

It is not new for science, mathematics, and engineering PhDs to enter careers in K-12 education or otherwise become involved with children's education. Most who became teachers, however, did so at private schools. By taking this path they avoided the onerous task of going back to college to earn the education credentials necessary for teaching in public schools. Thus, they often entered K-12 education without public school teaching credentials and usually without ever spending time in the classroom. Today, as the nation focuses anew on K-12 education, it is timely to consider ways to encourage PhDs to use their skills more widely in our schools.

The PhDs who are sprinkled around the nation's schools in teaching and other K-12 education positions often made their choice in spite of a system whose rewards seem to be reserved only for those who follow in the footsteps of their research advisers. Indeed, good researchers will always be needed, but that single-minded approach to a successful career can and should be changed. What is new about the National Postdoctoral Fellowship Program outlined in this report is that it draws attention to PhDs' interest in school careers by offering a path for using their talents to deepen K-12 education—a path that has the strong support of local, state, and national leaders. It demands collaboration and partnerships among academic institutions that produce teachers, the districts and state systems that oversee our schools, and our national science, mathematics, and education leadership. Its success will also depend on recognizing the diversity of our nation's schoolchildren and the centrality of that diversity in every level of implementation.

What are the expectations for such a program? As the committee rightly concluded, it will not solve the nation's teacher shortage or be a proper career choice for all PhDs. We hold high expectations for the fellows: We expect them to help improve and reform the K-12 learning environment, and in turn, must provide an exciting prospect for them as educators.

Education has always played a special role in this country. We have shared the belief that a good education leads children not only to have many choices as productive workers in the economy, but also to formulate their own meaning to life. A good education is the foundation for a lifetime of learning about both the natural and the engineered world, and it allows us to deal effectively with the problems associated with our everyday existence.

The fellows in the National Postdoctoral Fellowship Program will have a passion for discovery; they will respect the cultures of science, mathematics and technology, as well as the world of K-12 education; and they will be committed to achieving change through collaboration, while basing change on research on learning and teaching. This commitment represents an exciting challenge, and the implementation of this program will provide a new way for both scientists and society to view our nation's schools.

The committee's work would not have been possible without the help of individuals who shared their visions for what this project could accomplish and their insights on the intricacies of designing a program that first and foremost meets the needs of K-12 education. A number of colleagues helped with the writing, research, and editing and provided wise counsel at our meetings, including: Herman Alvarado, Amaliya Jurta, Judy Nyquist, and Stacey Piccirilli, National Research Council's Division of Policy and Global Affairs; Joe Alper, Consultant, Louisville, CO; Judith D'Amore, University of Washington; Ingrith Deyrup-Olsen, Professor of Zoology (emeritus), University of Washington; Margaret Hilton, Center for Education; Mary Grace Snyder, Coordinator, Resident Teacher Program, Montgomery County Public Schools; and Claudia Sturges, Director of Fellowship Programs, American Association for the Advancement of Science. We are also grateful for the guidance provided by Marilyn Baker and Charlotte Kuh of the National Research Council's Division of Policy and Global Affairs.

In the course of this study we held a workshop in Seattle, and we extend our gratitude to Dennis Schatz and the Pacific Science Center and Gerald M. Stokes and the Battelle Pacific Northwest National Laboratory for their hospitality. We also thankfully acknowledge the space provided for two other meetings at the Whiteley Center at the Friday Harbor Laboratories at the University of Washington and the University of Texas at Austin.

This report has been reviewed in draft form by individuals chosen for their diverse perspectives and technical expertise, in accordance with procedures approved by the NRC's Report Review Committee. The purpose of this independent review is to provide candid and critical comments that will assist the institution in making its published report as sound as possible and to ensure that the report meets institutional standards for objectivity, evidence, and responsiveness to the study charge. The review comments and draft manuscript remain confidential to protect the integrity of the deliberative process. We thank the following individuals for their review of this report: Joan Baratz-Snowden, American Federation of Teachers, Washington, DC; Liesl Chatman, Science and Health Education Partnership, University of California, San Francisco; Noah Finkelstein, Jacobs School of Engineering and Laboratory of Comparative Human Cognition, University of California, San Diego; Sherrie Hans, Consultant, Alexandria, Virginia; Toby M. Horn, DC Public Schools Office of Academic Services; Erin Peckol, BIOTECH Project, University of Arizona; Julie Strong, Menlo School, Athjerton, California; John D. Wiley, Office of the Chancellor, University of Wisconsin, Madison; and Lauren J. Young, Spencer Foundation, Chicago, Illinois.

Although the reviewers listed above have provided many constructive comments and suggestions, they were not asked to endorse the conclusions or recommendations nor did they see the final draft of the report before its release. The review of this report was overseen by Richard A. McCray, Department of Astrophysics, University of Colorado, and Henry Riecken, Professor of Behavioral Sciences, Emeritus, University of Pennsylvania. Appointed by the National Research Council, they were responsible for making certain that an independent examination of this report was carried out in accordance with institutional procedures and that all review comments were carefully considered. Responsibility for the final content of this report rests entirely with the authoring committee and the institution.

> M. Patricia Morse, *Chair*
> Committee on Attracting Science and
> Mathematics PhDs to K-12 Education

Executive Summary

The quality of science, mathematics, and technology education in kindergarten through high school (K-12) is a major concern for many observers of school systems across the United States. One key element in that concern has been the shortage of qualified teachers in these subjects. Significant percentages of the teachers of these subjects did not major or minor in them in college; some did not study them at all. The problem is particularly acute at the secondary school level in the nation's rural and urban areas. Moreover, because about two-thirds of the nation's K-12 teachers are expected to leave teaching in the next 10 years, the problem is likely to get worse. Other elements of concern have been the lack of adequate teacher professional development, curriculum development, and connections to educational resources outside the schools (including museums, aquaria, and zoos) and to institutions of higher education.

At the other end of the educational spectrum, the United States is justly famous for the quantity and quality of its doctoral graduates in these subjects. Yet an increasing number of well-trained PhDs cannot find—or decide they do not wish to pursue—traditional careers in academia or industry.

With interests that span these two ends of the educational spectrum in the United States, the National Research Council (NRC) has undertaken a three-phase project to explore the possibility of a program to attract science, mathematics and engineering PhDs to careers in K-12 education. The first phase of the project surveyed the interests of recent PhDs in science and mathematics in pursuing careers in secondary education. Analysis of the Phase I data suggests that a significant percentage of PhDs might be interested in pursuing careers in secondary education under some circumstances. This report from the second phase of the project presents a proposal for a national demonstration program to determine how one might prepare PhDs to be productive members of the K-12 education community. The proposed program is designed to help meet the needs of the nation's schools, while providing further career opportunities for recent PhDs in science, mathematics and engineering.

The committee proposes that the concept be demonstrated through a National Postdoctoral Fellowship Program to prepare new and recent PhDs for teaching and other positions in K-12 education and to prepare them to take part in future leadership activities. The program would provide 2 years of support for fellows to undertake classroom study and supervised teaching. Their work would include the courses and experiences necessary for teaching certification in their states. Part of the work would be based in institutions of higher education; part would be based in local schools. In the first year, all costs would be borne by the national program; the schools in which the fellows are employed are expected to pay their stipends and benefits in their second year in the program.

The proposed demonstration program needs to be national in scope because the needs in K-12 education are national, the potential supply of PhDs represents a national pool, and there would be economies of scale for recruitment, selection, and placement at the national level. A national program will also provide the opportunity for schools to choose from a larger pool of applicants and for applicants to choose from a larger pool of schools than would be the case for local or state programs. Finally, the committee believes that the nature of the problem is such that success will require the prestige and momentum that can only be achieved through national attention.

The committee strongly endorses Phase III of the project in which the proposed demonstration program would be implemented and evaluated. An initial 4-year demonstration program for cohorts of around 15 fellows per year should generate enough evidence to evaluate the feasibility and desirability of expanding the effort. However, the committee believes that a comprehensive evaluation of whether the fellows become successful K-12 educators and improve K-12 teaching in science, mathematics, and technology in the nation's schools will require a program of at least 30 fellows a year for perhaps 10 years. For 30 fellows per year, such a program would cost about $2 million for the first year, with all support being provided at the national level, and about $500,000 the second year, when most of the support would be provided by the schools in which the fellows are working. Thus, at steady state, the National Postdoctoral Fellowship Program suggested here would require about $2.5 million (in 2002 dollars) for 30 fellows; for a program of 60 fellows, the estimated cost would be about $4.8 million. These estimates include a basic annual stipend ($35,000) and benefits ($15,000) for each fellow in the first year, and some administrative costs in both years; but they do not include ancillary costs associated with evaluation and administrative overhead.

The committee recognizes that the details of the demonstration program will have to be specified in Phase III and will depend in part on the source of support for the program. The committee believes that one or more federal agencies, such as the National Science Foundation, the National Institutes of Health, and the Department of Education, are the most likely funders for this program because of its relevance to their responsibilities; however, its support might also come from private foundations.

1

Introduction

Education has long been both a source of pride and strength for the United States and a topic of concern. The support of free public schools for all children in each state was an early demonstration of the nation's belief in and commitment to education. Criticism of those schools and the quality of that education has been a part of the nation's dialogue for almost as long. As the twenty-first century begins, much of the concern about the nation's education is focused on the quality of science, mathematics, and technology instruction and learning from kindergarten through high school (K-12). This report proposes a national demonstration program to help address that concern.

A NEED AND AN OPPORTUNITY

In a world in which science and technology play an ever-increasing role in determining a nation's economic and social health, U.S. children are falling behind their international colleagues. Results from the Third International Mathematics and Science Study (TIMSS) in 1995 showed that, while U.S. fourth-graders scored above the international average in mathematics, eighth-graders scored slightly below the average and high school students scored near the bottom in both science and mathematics compared with their counterparts in other countries (National Center for Educational Statistics (NCES), 1999a). When the TIMSS was repeated in 1999, U.S. students' performance compared to 1995 was unchanged, aside from a significant increase in the performance of eighth-grade black students in mathematics (NCES, 2000).

Similar information comes from the National Assessment of Educational Progress (NAEP). In the 1996 NAEP, less than one-third of all U.S. students in grades 4, 8, and 12 performed at or above the "proficient" achievement level in mathematics (Reese et al., 1997) or science (National Assessment Governing Board (NAGB), 2001). Although the recently released NAEP 2000 results for mathematics indicate that performance of students in grades 4 and 8 has improved significantly over the last decade (the results for grade 12 were mixed), once again less than a third of all students performed at or above the "proficient" achievement level (U.S. Department of Education, 2001). In science, the results "show no significant changes in grades 4 and 8, and a decline in performance at grade 12 since 1996" (U.S. Department of Education, 2002).

Four decades after Presidents Eisenhower and Kennedy made science education their top priority for American education and a decade after President George H.W. Bush, the nation's governors, and business leaders declared, "[b]y the Year 2000 . . . United States students will be first in the world in mathematics and science achievement" (U.S. Department of Education, 2000), the country is still struggling to find ways to lead children to perform at high levels in science and mathematics. The lackluster performance of U.S. students in science and mathematics has many causes, and it is not the purpose of this report to review them save for two: (1) K-12 science, mathematics, and technology education suffers from critical shortages of qualified teachers, and (2) there is inadequate support for teachers' professional development, for effective curriculum development, and for strengthening connections between K-12 education and higher or informal education.

At the other end of the spectrum of U.S. education, the nation has an outstanding record in producing some of the world's most renowned scientists, mathematicians, and engineers, as well as a large number of highly qualified and productive doctorates. Yet several recent studies indicate that the number of postdoctoral fellows in the United States has been growing and that they, as well as PhDs who do not undertake postdoctoral work, have experienced difficulty in finding permanent careers in either academia or industry (National Research Council (NRC), 1998, 2000c). Although the situation is most pronounced in the biological sciences, significant numbers of PhDs in other scientific fields also report disappointment in their prospects for finding careers as university faculty (NRC, 2000c).

This mismatch between the number of PhDs being produced and the number of faculty and industry positions for them is causing PhDs in all fields to reconsider their career options (Fiske, 2001; Kreeger, 1998; NRC, 1995, 1996a; Rosen and Paul, 1997; Robbins-Roth, 1998; Tobias et al., 1995; Nyquist and Woodford, 2000; Nyquist, 1999). Moreover, results from Phase I of this project suggest that a significant number of PhDs might be attracted to careers in K-12 education under some conditions (see the discussion in the next section). This situation affords the nation a golden opportunity to consider applying the talents of these PhDs to address an important need facing the nation's schools—improving K-12 science, mathematics, and technology education.

In a recent editorial in the *Chronicle of Higher Education,* Vartan Gregorian, president of the Carnegie Corporation of New York, articulated an idea for rebuilding the nation's public school systems around a revitalized teaching profession. That editorial included the following vision of this opportunity:

> [W]e should remove artificial distinctions among teachers. The idea that Ph.D.'s belong in higher education and M.A.'s and B.A.'s belong in elementary and secondary schools does not make sense. It should not be considered a step down for a Ph.D. to teach in a public high school. . . . Colleges should urge Ph.D's to consider teaching in our high schools, and high schools should welcome them. Encouraging more Ph.D.'s to become teachers in public schools would greatly enrich those programs. (Gregorian, 2001, p. B7)

THE COMMITTEE'S TASK

This report covers the second phase of a three-phase project to explore the possibilities and design of a demonstration program to attract PhDs in science, mathematics, and engineering to careers in K-12 education. The goal of this project is to create a cadre of educators who are deeply knowledgeable about science, mathematics, and engineering and deeply committed to K-12 education—who will be capable of helping children learn and understand these key aspects of the modern world.

Phase I of the project undertook a major national survey of science and mathematics graduate students and postdoctoral fellows to ascertain the conditions under which they would be interested in teaching careers at the secondary school level. (See Appendix A for the executive summary of the Phase I report.) This report marks the completion of Phase II. Its aim is to inform policy makers at both the state and federal levels of ways to attract and retain PhDs in K-12 education as well as to present frameworks for the design and evaluation of demonstration programs to accomplish that goal. The charge to the Phase I committee directed it to consider only issues in secondary education; the charge to this committee explicitly includes all of K-12 education. The committee agrees that the evidence it examined favors a demonstration program at the secondary level, and that it would be appropriate to focus the initial demonstration at that level. However, the committee also believes that science, mathematics and engineering PhDs could make important contributions to primary education. Given that conviction and the committee's charge, the scope of this report is K-12, rather than "secondary" education.

In the Phase I survey, 36 percent of respondents expressed interest in careers in secondary education. The survey also asked "Would you consider other full-time positions in K-12 science and mathematics education?" Although few of the respondents expressed an interest in becoming K-6 teachers, between 55 and 82 percent of the respondents interested in secondary school teaching positions indicated that they were interested in becoming science or mathematics specialists for a school district, science resource teacher for a school; or taking positions in curriculum development, teacher professional development, science education partnerships, university-based science education research centers, and science museums or similar institutions (NRC, 2000a). With this preliminary evidence in hand, the National Research Council NRC broadened the scope for Phase II to include careers in K-12 education other than secondary school teaching. In response to a suggestion from the NRC Governing Board Executive Committee, the committee also broadened the scope to include PhDs in engineering.

As part of its charge, the committee was directed to hold a public workshop to examine the results from the Phase I study and to use those data to develop plans for demonstration projects. That workshop was held on June 2-4, 2000, in Seattle, Washington. The workshop brought together PhDs in science, mathematics, and engineering with careers in K-12 education; state policy makers; teachers; teacher educators; as well as representatives of teachers' professional organizations and unions, federal and private funding agencies, and science and mathematics professional societies. They and the members of the committee discussed the key issues in

INTRODUCTION

considering a demonstration program to recruit PhDs in science, mathematics, and engineering to work in K-12 education including:

- needs in the K-12 system that might be addressed by such a program;
- specific K-12 education careers that might be open to program participants;
- what further preparation PhDs would need for successful careers in K-12 education;
- recruitment of local program providers and of Ph.D. applicants; and
- potential sources of funding for various components of the program.

The agenda and participant list for the workshop are in Appendix B.

In addition to the valuable information gained at the workshop, the committee used various sources of data and information in its work, including the national consensus standards for K-12 science, mathematics, and technology education; previous NRC reports on issues of teacher supply and the improvement of science, mathematics, and technology education in grades K-12; and the general research literature on K-12 education, graduate education, and postdoctoral experiences. It also relied heavily on the results of the Phase I part of the project. Nevertheless, designing a demonstration program like the one proposed in this report involves many factors for which the empirical research base is only suggestive. Thus, the committee also relied on the experiences of its members and other experts in relevant organizations, along with the contributions of the workshop participants. In this respect, it is similar to a recent report that lays out the design for a new approach to education research (NRC, 1999d).

Chapter 2 of this report discusses the need for qualified teachers and improved teacher-support structures in the nation's schools, focusing on science, mathematics, and technology subjects. Chapter 3 presents information on young PhDs' interest and retention in K-12 education. Chapter 4 then presents the committee's proposal for a national demonstration program to attract and retain PhDs in K-12 education.

2

The Challenges of Education

To be effective, professionals in K-12 mathematics and science education must have a deep understanding of two cultures—the world of inquiry and problem solving that is central to mathematics and science and the world of facilitating learning in the classroom. This understanding involves three areas: subject-matter or content knowledge, pedagogical content knowledge, and pedagogy (National Research Council (NRC), 1999b). That is, high-quality teachers have expertise in the subject matter they are teaching, in how to teach, and in how to teach a specific subject—teaching English is not the same as teaching mathematics.

THE NEED FOR HIGH-QUALITY TEACHERS

The United States now faces a shortage of teachers, especially of qualified teachers. That shortage has resulted in the hiring of uncertified or underqualified teachers. According to the 1996 study by the National Center on Teaching and America's Future (NCTAF) (1996), more than 50,000 inadequately prepared teachers enter the profession each year. Just last year, the *New York Times* (July 1, 2001) reported that 60 percent of the teachers hired for New York City were uncertified (Goodnough, 2001). And the problem is not likely to be resolved in the foreseeable future: with some two-thirds of the nation's K-12 teachers expected to retire or leave the profession over the coming decade, the nation's schools will need to fill between 1.7 million and 2.7 million new teaching positions (National Center for Education Statistics (NCES), 1999a).

Of those new teaching positions, about 200,000 will be in secondary science and mathematics (NRC, 2000a). In the face of the current shortages of qualified teachers, there is no reason to expect that a significant percentage of the people hired for these positions will be proficient in both subject-matter background knowledge and pedagogy in these subject areas. Among high school physics teachers, for example, 32 percent have a degree in physics or physics education and have taught it on a regular basis, 41 percent have no physics degree, but have extensive physics teaching experience, and 27 percent have no physics degree and little physics teaching experience (American Institute of Physics, 1999). Another report (Ingersoll, 1999) indicates that approximately 33 percent of mathematics teachers and 20 percent of science teachers in grades 7-12 do not have either a major or minor in their field. These underqualified teachers teach more than 26 percent of mathematics students and more than 16 percent of science students. In urban and small rural school systems, especially those with large populations of students in poverty, the percentages of underqualified teachers are even higher.

Having well-prepared teachers is central for students' becoming literate in science, mathematics, and technology. A report by the NCTAF (1996) unequivocally shows the positive effect of better teaching on student learning. Another study, by the Center for the Study of Teaching (Darling-Hammond, 1999), found that the two most consistent and powerful predictors of student achievement in science and mathematics were having teachers who were both fully certified and had a college major in the subject being taught. These findings about the importance of qualified teachers are consistent with research on what experts know and how they can use that knowledge. Teachers with content expertise, like experts in all fields, understand the structure of their disciplines; they thus have cognitive "roadmaps" to guide the assignments they give students, the assessments they use to gauge student progress, and the questions they ask in the give and take of the classroom (NRC, 1999b).

Teachers face a further challenge in the classrooms of the twenty-first century: the increasing diversity of the nation's schoolchildren. The waves of immigration to the United States in the last decades of the twentieth century have filled the schools with children from myriad cultural and ethnic backgrounds; these students have varying degrees of English proficiency when they begin school. Fairfax County, Virginia, a suburb of Washington, DC, is the oft-cited example: the children in the school system come from more than 180 different countries and live in homes in which over 100 languages are spoken. To meet the goal of a good education for *all* children, teachers must have the training and expertise to understand the diverse group of students in their classrooms.

NEW APPROACHES FOR PROFESSIONAL DEVELOPMENT

Although there are many exemplary teachers in the nation's public and private schools, many other teachers, especially in primary and middle schools, have not received adequate preparation in science and mathematics. For current—and future—teachers who arrive in classrooms with inadequate training, professional development on the job can be a key route to achieving more knowledge and better skills. Particularly in the context of evolving national and state standards in science, mathematics, and technology (American Association for the Advancement of Science (AAAS), 1993; NRC, 1996b; National Council of Teachers of Mathematics, 1989, 2000; International Technology Association, 2000; Eisenhower National Clearinghouse, 2001a), there is an urgent need for both professional development and new instructional materials and curriculum related to science, mathematics, and technology. These standards and related documents (AAAS, 1990; Hurd, 1997; Krajcik, 1999; Alberts, 2000; Minstrell, 2000; NRC, 2000d) call for greater use of teaching through inquiry, problem solving and design. They also call for the teaching of science, mathematics, and technology in a contemporary context that is relevant to students' lives and experiences.

The trend towards using hands-on, inquiry-based activities to teach primary grade-level science and mathematics to students takes advantage of the fact that young children seem predisposed to learn about their world, and it can make learning exciting to teachers and students alike (National Science Resources Center [NSRC], 1997). But most teachers need assistance to implement these new approaches. One source of support for teachers are science, mathematics, and technology supervisors, who serve as curriculum experts, set up resource centers, and provide technical assistance for integrating these hands-on teaching approaches—as well as other learning tools such as computer hardware and software—in K-12 classrooms. Many school districts have such supervisors, who also often help to write grants and pursue other avenues for obtaining resources for schools. These supervisors, like the teachers they support, need to be well qualified in the content of science, mathematics, and technology education as well as in pedagogy.

The professional development needs of teachers are the focus of various national efforts. The report of the National Commission on Mathematics and Science Teaching for the 21st Century (2000) (also known as the Glenn Commission, for its chair) called for the nation to establish intensive Mathematics and Science Summer Institutes to provide in-depth professional development to science, mathematics, and technology educators so that they can reach their full potential as teachers. The National Science Foundation (NSF) is funding Centers for Learning and Teaching (see NSF 2001c), Centers for Advanced Technology Education (see NSF, 2001d), and Mathematics and Science Implementation Centers (see NSF, 2001e) to provide familiarity and implementation assistance with state-of-the-art classroom materials for teachers of science and mathematics. Moreover, a major activity of the Department of Education's ENC for science and mathematics instructional materials is to get useful education tools into the hands of science, mathematics, and technology teachers (see ENC, 2001b). Other new programs, such as the capacity-building grants funded by the Department of Education, the NSF, and the National Institutes of Health (NIH) through the Interagency Education Research Initiative, call for more thorough dissemination of effective standards-based science, mathematics, and technology instructional materials and for the teacher professional development needed for their implementation (see NSF, 2001f).

Activities for teachers' professional development are changing to reflect growing knowledge about the requirements for improved and effective teaching (see NCTM, 2000). In the past, many professional development activities were based on summer institutes, publisher-provided enhancement for use of their textbooks and other instructional material, and university-based professional development courses. New activities for ongoing professional development for teachers focus more on content updates, experience with exemplary instructional materials, and practice with assessments that are connected to the educational goals of the curriculum. One approach that is in its infancy is support for teachers for activities in the classroom—perhaps, through establishing electronic learning communities among teachers, scientists, mathematicians, and engineers. The NSF has recently funded a series of such implementation centers—a series of Centers for Learning and Teaching—and has plans to expand this approach in the years ahead (see NSF, 2000).

In addition to institution- and curriculum-based professional development activities, there has been a significant increase in number and quality of community-based centers and activities for informal education, particularly in the last 15 years. For example, the number of open staff positions among members of the Association of Science and Technology Centers (ASTC) increased from about one per year prior to 1975, to about seven per year between 1975 and 1984, to more than ten per year between 1985 and 2001 (ASTC, 2001). These positions include jobs in science, technology, and natural history museums; botanical gardens; aquaria; and environmental centers associated with national, regional, and local parks and nature preserves. These centers have created new and expanded opportunities for professional development for teachers, participation by students in scientific research, and enhanced connections among teachers, students, and scientists. This interface has more recently become a magnet for scientists who see the value of good science for K-12 students. Such centers provide the opportunity to connect researchers with teachers and to promote active learning for teachers and students. The researchers are in a position to convey the excitement of science and provide resources for teachers' follow-up activities in the classroom. The teachers bring understanding of new approaches to pedagogy and of how students learn.

Schools are also turning to nontraditional sources of help for professional development and teaching materials. New partnerships are being forged among scientists, science educators, and teachers. These partnerships, often facilitated by the use of computers and scientific databases, may be connected with industry, academic institutions, or informal education sites such as museums (see Cohen, 1997; Munn et al., 1999; Sussman, 1993). Some partnerships are developing innovative instructional materials that are disseminated free to teachers, such as those of the Howard Hughes Medical Institute (2001) and the American Chemical Society (2001).

Local science centers, museums, parks, zoos, botanical gardens, and aquaria are using their in-house resources to develop standards-based activities and teaching materials that can supplement standards-based lesson plans. In many instances, industry, academia and these informal educational settings are also partnering with schools to provide teachers' opportunities for professional development during summers (NRC, 1996c). Many government agencies with science, mathematics and technology missions are developing K-12 educational materials, as well as providing related professional development for teachers (see, for example, NIH, 2001; National Aeronautics and Space Administration (NASA), 2001; U.S. Department of Agriculture (USDA), 2001; and U.S. Department of Energy, 2001). In less structured ways as well, scientists, mathematicians, and engineers at local companies often serve as accessible sources of information for teachers.

BEYOND PROFESSIONAL DEVELOPMENT

It is clear that many individuals, organizations, and institutions are developing the resources to enable K-12 science, mathematics, and technology teachers to increase their own knowledge and improve their students' learning. But it is equally clear that individual programs—even very good ones—scattered across the nation's more than 16,000 school districts cannot meet the nation's need for high-quality science, mathematics, and technology education in K-12 classrooms. And even with very good materials and opportunities for professional development, follow-through and continued support for teachers in the classroom are often missing (see NRC, 1999c). A critical way to help provide that support is to build many more direct bridges between the practitioners of science, mathematics, and engineering and K-12 schools, teachers, and students. These bridges can be encouraged and maintained by employing selected individuals in school districts who deeply understand both the culture of science and the very different culture of the schools. On the other side of the divide, similar types of individuals are needed inside institutions rich in science and technology to facilitate the connection of these institutions to school districts and schools. As schematically indicated in Figure 2-1, a few such individuals, strategically placed, can thereby catalyze an enormous increase in the personal contacts and resource flows between the K-12 education system and the vigorous scientific and engineering community in the United States.

In summary, there is an urgent need for more permanent connections between two very different worlds: that of science and that of the schools. This need provides one important basis for the committee's proposed national demonstration program. Before turning to that proposal, in Chapter 4, the next chapter considers the people who could in principle provide the strongest links between the world of science, mathematics, and engineering and the world of schools, teachers, and students.

Science-rich institution
(university, museum, etc.)　　　School district

Strong interactions
between the Fellows

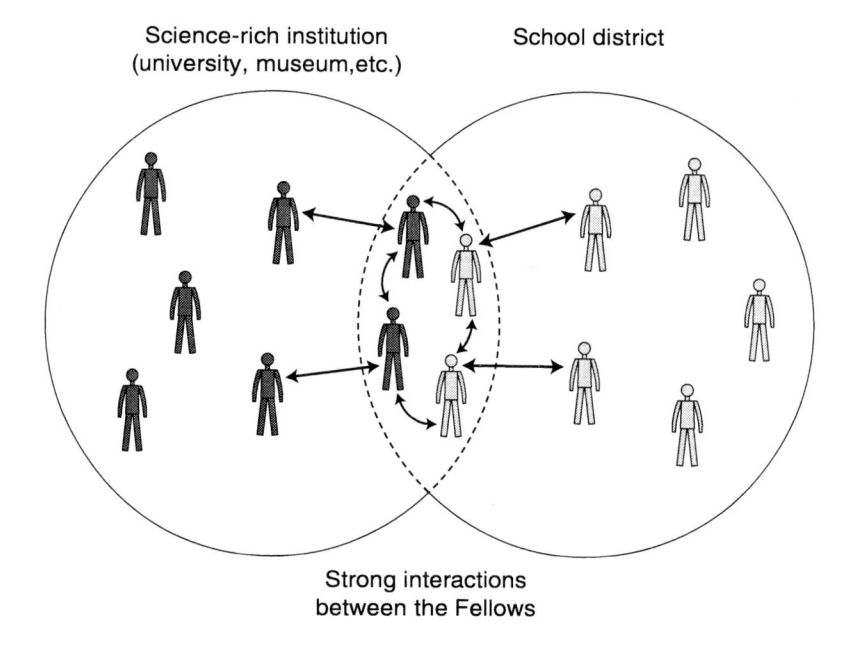

FIGURE 2-1 Why PhDs trained in the proposed program can serve as ideal connectors between science-rich institutions and schools.

Every institution has its own culture, which makes it relatively easy for individuals inside of the same institution to interact. But, it creates serious difficulties for making meaningful connections between two institutions whose cultures are as different as a university science department and a school district. After their training, the PhDs discussed in this report (placed schematically between the *dotted lines*), share values that allow them to form strong personal bridges between two such institutions. Ideally, some would be employed in school districts and others in science-rich institutions, allowing them to catalyze a broad exchange of information and resources.

3

From Grad School to Grade School

The premise of this National Research Council (NRC) project is that recent PhDs may provide a source of highly qualified professionals for K-12 education in science, mathematics, and technology. Why focus on people who have science, mathematics, and engineering PhDs? Two of their obvious attributes are deep content knowledge, which is a requirement for good teaching, and a commitment to and interest in teaching, which is often the goal of PhDs, though usually at the college and university levels. This chapter considers in greater depth what are known about PhDs and about the potential of their pursuing careers in K-12 education. It looks at three questions: Are they interested? What do they know that is special? What do they need to learn? It ends with brief looks at the questions: How can they learn what they need to know? And how can they be retained in K-12 education?

ARE THEY INTERESTED?

The first phase of the project on attracting PhDs to K-12 education was devoted to investigating whether recent PhDs might consider careers in K-12 education and under what conditions they would do so. In carrying out this second phase of the project, the committee relied heavily on the work of the Phase I committee (NRC, 2000a; see Appendix A for the report's executive summary) and on the discussions at a workshop (see Appendix B for the agenda and list of participants).

The Phase I committee investigated the career ambitions of recent and prospective PhDs in the physical sciences, life sciences, and mathematics, and their interest in taking positions in secondary science, mathematics, and technology education under a variety of hypothetical conditions. Through focus groups and a national survey of more than 700 graduate students and postdoctoral fellows, the committee asked respondents how teacher preparation programs, work conditions, and compensation packages could be modified to attract them to careers in secondary and primary school education.

Respondents to the survey had typically considered at least four different options in contemplating their careers; 36 percent of them said they had considered secondary school teaching or other secondary education positions. This number is significant, because, according to a special tabulation of data from the 1997 Survey of Earned Doctorates that was conducted for the Phase I report, only 0.8 percent of all PhDs currently work in K-12 educational institutions (NRC, 2000a).[1]

The range of positions for which the survey respondents expressed interest included not only becoming a science or mathematics teacher, but also becoming a science or mathematics specialist for a school district, in working in a university- or industry-based science educational partnership, or in serving as a science specialist in a science resource center. Some respondents also expressed interest in working on curriculum development or with education programs of a science museum, environmental science center, or similar type of institutions.

A key question that arose in the Phase I committee's work was why less than 1 percent of PhDs are currently working in K-12 educational institutions given that 36 percent of recent PhDs have considered secondary teaching? One answer is that there are many negative perceptions about K-12 teaching that work against entering this career. Those negative perceptions include:

- a lack of status and respect for teachers,

---

[1]Respondents who were still in graduate school, female, or U.S. citizens were more likely to say they had considered careers in secondary education than respondents who were postdoctoral fellows, male or non-U.S. citizens. Respondents in chemistry, with strong career options in industry, were less likely than respondents in the biological sciences, physics, and mathematics to consider secondary teaching positions.

- poor classroom laboratory facilities,
- too many students per class and student discipline problems,
- structured curricula that allow little room for creativity,
- possible conflicts with teachers who do not hold PhDs, and
- low salaries.

There are also negative perceptions from school administrators and university faculty that may act as disincentives to prospective PhD teachers. For school administrators, there is the belief that PhDs may have good content knowledge but do not know how to teach or to relate to K-12 students. For university faculty, the prevailing perception was well captured in the Phase I report:

> [M]any university faculty do not promote nonacademic careers for PhDs, much less careers in secondary school education. Indeed, graduate students typically aspire to positions in academic science and mathematics similar to those of their mentors, and the socialization process in graduate school strongly reinforces this career path (NRC, 2000a, p. 4).

Given the many negative perceptions that may discourage PhDs from pursuing careers in K-12 education, perhaps it is surprising that more than one-third of graduate students and postdoctoral fellows have considered K-12 teaching. The Phase I committee noted that its work also revealed some attractive features of a K-12 teaching career:

> While many participants in our focus groups held negative perceptions, many also held a number of positive perceptions, which included attractive working hours, a work schedule similar to their children's school day, and time for research or other activities during the summer. Many also believed they would enjoy the opportunity to foster the scientific interests of young people (NRC, 2000a, p. 4).

During the workshop held in connection with this report, the participants with PhDs who work in K-12 education provided additional insights as to why they are so few in number. They stated that a major barrier for anyone leaving academic research is the potential loss of respect from that community. Although they were ultimately happy with their decisions to enter careers in K-12 education, they suggested that this barrier could be lowered if the PhDs were offered a prestigious national fellowship by an institution that is respected by the academic research community.

This finding agrees with what the Phase I committee found in its survey, when it explored under what scenarios the survey respondents would consider secondary school teaching. The item that received the largest number of positive responses, checked affirmatively by 67 percent of the respondents, was the award of a prestigious national fellowship that provides training and placement, and covers living expenses. Some other items that solicited significantly positive responses were the availability of mentors, access to regional or university-based science teaching resource centers, and field opportunities for research that might involve their students. A requirement to undergo a normal full teacher certification process was a negative factor to survey respondents, but they were quite amenable to an accelerated program.

The Phase I committee concluded that large enough numbers of PhDs appear to be sufficiently interested in secondary education for the NRC to continue the next phase of the project. This committee concurs with that conclusion and so turns to the next question.

WHAT DO THEY KNOW?

Perhaps the most important attribute that PhDs would bring to K-12 classrooms is a deep content knowledge of science, mathematics, or technology, including extensive experience with science as inquiry and similar forms of scholarship. As discussed in Chapter 2, all of the national standards for education call for greater use of teaching through inquiry, problem solving, and design. Moreover, a deep understanding of the content of mathematics or science is critical for high-quality teaching (NRC, 2000b), and it would also be critical for other careers in K-12 education.

A second key attribute of PhDs is a passion for science and mathematics: that passion could help inspire both their teaching colleagues, schools and their students. With appropriate preparation (see next section), they should be able to harness their enthusiasm for discovery to teaching and to the development and implementation of innovative programs in science, mathematics, and technology.

Along with their passion for inquiry and their insights into science, mathematics, and technology, PhDs—especially recent PhDs—are also intimately familiar with the use of information technology in their work.

This familiarity with modern tools of inquiry is a third attribute that PhDs would bring to K-12 education. In many cases, they have become expert in using the Internet for library work, for acquiring data for research (from databases, remote sensing devices, and the like), for collaborative work with others, and for the analysis and the other everyday tasks of their work. Their ability to use technology can be helpful in solving problems and developing innovations in K-12 instruction, and it can also serve as a model for learning for their students.

In addition to the above attributes, science, mathematics, and engineering PhDs will have a strong connection to at least one institution of higher education—the university at which they earned a degree. They can use those ties to build strong partnerships between K-12 and higher education.[2] In addition, their ties with national professional societies can provide another source of connections between K-12 education and the broader world of science, mathematics, and technology.

WHAT DO THEY NEED TO LEARN?

Although PhDs have a deep content knowledge in their fields, they are likely to lack the two other ingredients necessary for high-quality teaching: pedagogical knowledge and pedagogical content knowledge. That is, they probably do not have the general skills of a teacher, nor do they know specifically how to teach particular science, mathematics, and technology subjects in ways that enable most of their students to learn. They also know little about the opportunities and limitations imposed by the structure of the school day and other classroom realities. In fact, much of the training for a PhD is in some ways antithetical to the training needed for K-12 teaching.

Most PhDs have gone through a demanding program that is long, difficult, and that requires a narrow concentration on problems near the cutting edge of their subfield. It is unlikely that the last few years of their apprenticeship will have provided an opportunity for exploring a broad interest in their field—or for learning about science in general, or how to teach science. The PhDs who become K-12 teachers will need substantial

---

[2]Legislation has been introduced in Congress to provide funding to establish partnerships between K-12 schools and institutions of higher education.

classroom teaching experience, and they must develop an understanding of the empirical and theoretical research findings about teaching and learning.

Pedagogical Content Knowledge Research shows that the content to be taught must be appropriate for the age and level of cognitive development of K-12 students. Teachers need to understand the effects of students' prior knowledge and their misconceptions on their ability to learn new concepts (NRC, 1999b); they must also know how each type of misconception can be addressed to increase student understanding (Minstrell and van Zee, 2000). Teachers must develop an understanding of how children make sense of a discipline through the learning experiences and instructional materials used in quality K-12 mathematics and science programs (NRC, 1999b). They also need to understand how the structure of science and mathematics programs is used to allow students to grow in their mastery of both content and process throughout their K-12 years. Finally, national and, more recently, individual state content standards and curriculum frameworks are increasingly important in determining what is taught in U.S. schools, and teachers need to learn what they are and how to make effective use of them in the classroom. This entire area of knowledge, called pedagogical content knowledge, is one that most recent PhDs in science, mathematics, and engineering will need to learn.

Pedagogy The learning sciences have made great progress in the past decade in understanding how children learn (NRC, 1999b). New and recent PhDs in science, mathematics, and engineering are unlikely to be familiar with what the learning sciences have contributed to this field of cognition. And yet our understanding of how people learn provides the conceptual structure for strategies and techniques that teachers use to help children understand science and mathematics. The learning sciences also provide the basis for designing effective assessments of student learning and knowledge (NRC, 2001a). The PhDs who wish to move to K-12 education will thus need to learn about the theories of learning and the experiments that support them; how our knowledge of cognition relates to teaching based on scientific inquiry or mathematical problem solving, as called for in the national education standards; and how this relates to assessing student learning appropriately.

Context and Diversity The demographics in the United States have undergone rapid change in recent decades, and that change is reflected in the nations' classrooms. An increasingly large number of the nation's children are either immigrants or the children of recent immigrants, and they bring a wide variety of perspectives, expectations, and experiences to school. Research shows that context, including ethnicity and culture, plays a large role in how people learn (NRC, 1999b). PhDs who are moving into K-12 teaching will need to understand what effect these differences have on students, both inside and outside the classroom. When students in a classroom have a variety of cultural and ethnic backgrounds, their teachers need to use a variety of approaches in teaching and learning. They will need to know the effects on children who feel marginalized by their color or ethnicity, how tracking in school programs affects students from different backgrounds, and, more broadly, how class and race stratification issues affect American education.

Classroom Management Every K-12 teacher needs to know how to manage classrooms in ways that promote learning. PhDs who may have taught college-level courses will probably need some training to manage classrooms full of teenagers or younger students. Some particular topics that need to be covered are how to maintain discipline in classrooms, plan lessons that engage students, keep records, organize classrooms, and distribute instructional time. To become teachers, they will also need to know how to manage and distribute to their students the resources for the inquiry and experiential laboratory-based learning that are called for in the science, mathematics, and technology national education standards.

Schooling To be effective in the K-12 environment, teachers need to understand the challenges and opportunities facing schools and how the schools respond to them. In particular, they need to have some understanding of district and local school administration; education policy at both the state and federal levels; issues related to special populations; sources of funding for special programs; the influence of parents and policy makers on K-12 education; and liability issues and accountability—including student assessment, program assessment, and reporting.

HOW CAN THEY LEARN WHAT THEY NEED TO KNOW?

In thinking about an appropriate program to teach PhDs what they will need to know to be high-quality teachers, it is useful to examine examples of other programs for nontraditional pathways to K-12 education.

According to Haury (1998, p. 4), teacher education programs for midcareer or second career professionals "may, in fact, seem the same as traditional teacher preparation programs in terms of course requirements and field experiences," but a major difference is that the courses take the experiences of the candidates into account. The courses are often taught at the graduate level, even for those teacher candidates who do not hold PhDs. Because some topics can be skipped or covered at an accelerated pace, the programs for nontraditional students tend to be much shorter than the standard 4-year program for undergraduates. Appendix C describes three such programs: UTeach at the University of Texas at Austin; the midcareer mathematics and science program at the Harvard Graduate School of Education; and Teach for America, which was begun as a senior thesis project at Princeton University. Although none of these is designed for PhDs in science, mathematics, and engineering, they provide examples of the kinds of courses, orientation, and classroom experience that might be appropriate for people moving from graduate school to K-12 teaching.

Other sources of information for preparing PhDs for K-12 classrooms are available from teacher professional organizations, science and mathematics professional societies, and educational researchers. These groups have all paid a great deal of attention to the preparation of teachers over the last decade. Two key compilations of these ideas and the results of research are the report from the National Academies on the education of teachers (NRC, 2000b), and the report of the Glenn commission (National Commission on Mathematics and Science Teaching for the 21st Century, 2000).

Although the amount of material that PhDs need to learn in order to teach well may seem daunting, it is surely less daunting than the challenges they have overcome in completing their doctorates. And they will bring to this learning task their deep content knowledge; their extensive experience as learners; and a commitment to teach and communicate to K-12 students their interest in, knowledge of, and passion for science, mathematics, and technology.

HOW CAN THEY BE RETAINED IN K-12 EDUCATION?

The committee was unable to identify unique retention strategies for keeping PhDs in K-12 education careers. One aspect of the demonstration program might be to determine whether some unusual strategies can be developed. Retention strategies that have been shown to be effective for teachers include: one-on-one mentoring, with mandatory participation for all new teachers, involving teachers in decision making, implementing team or interdisciplinary teaching, and making scheduling changes to allow common planning time for teachers in schools. (National Center for Teaching and America's Future, 1996; Hare and Heap, 2001; National Association of State Boards of Education, 1998). Retention strategies for the broad array of professionals in other aspects of the K-12 education system have not been a major policy concern, and there is a paucity of information from which the committee could address retention in these other types of careers.

4

Proposed Demonstration Program

This final chapter turns to the design of a demonstration program for new and recent PhDs in science, mathematics, and engineering. The committee's proposed demonstration program is based on what was learned from Phase I of this project about the attitudes of the candidates for this program (see Appendix A), on the discussion in Chapter 3 of what prospective teachers need to know, on a few related programs that have been launched in recent years, and on the committee members' judgment and experience.

This demonstration program is not proposed as a step on the path to *the* solution to the current and expected continuing shortage of highly qualified teachers in science, mathematics, and technology in K-12 classrooms. Rather, it is offered as one potential contribution toward a broad approach to improving the quality of K-12 education in these subjects (see Appendix D for a description of some related programs). And it is also offered to generate an option for new and recent PhDs who may find the challenges and rewards of K-12 teaching more attractive than more traditional academic or industrial careers.

The overarching goals of the proposed demonstration program are to work out ways of preparing new and recent PhDs to become certified K-12 teachers, so that they can use their unique set of skills to significantly improve and transform K-12 teaching and learning in science, mathematics, and technology. By marrying the attributes of PhDs with the pedagogical and other teaching skills they acquire in the program, the nation will gain a group of K-12 professionals who possess deep content knowledge, a mastery of inquiry and analytic skills, direct connections to current research and laboratories, an ability to engage students and K-12 teacher colleagues in original research, and knowledge of the culture of science. These professionals can serve as role models for students who might want to become scientists or engineers. Very importantly, they can bridge the very different cultures of science and the schools in a variety of ways. They have the potential to form broad connections between universities and school districts; to become leaders in science and education at the district, state, and national levels; and to be advocates for school systems to gain access to resources at museums, zoos, aquaria, industries, and universities.

A key to the success of the demonstration program proposed in this report is designing it to be responsive to the needs and interests of the PhD fellows, to the institutions that will conduct the teacher education, and to the local communities in which they will ultimately work. The latter two groups must be responsible for the operation and administration of the preparation programs; they will also serve as a "home" for the fellows.

Four features critical to the demonstration program are that it:

- is national in scope;
- is 2 years in length;
- provides financial and other support for participants; and
- is designed to provide the opportunity for the participants to obtain state teaching certification.

The next section discusses these four program features; this will be followed by a discussion of other program features that will be important for the program's success: recruitment, selection and placement, teacher preparation, and mentoring and leadership preparation. The final sections briefly discuss the demonstration program's structure, funding, and evaluation, closing with a brief look at next steps.

A NATIONAL 2-YEAR FELLOWSHIP

A National Program

We propose a demonstration program that is national in scope for four reasons: (1) the national needs in K-12 science, mathematics, and technology education; (2) the mobility of PhDs; (3) a desire to offer the maximum range of choices and opportunities for the participants and schools; and (4) a belief that the issues and ideas embodied in this demonstration program deserve national attention.

People with new and recent PhDs in science, mathematics, engineering, and technology are not evenly distributed among the states, and there are only a few places that might have enough of them to mount a viable demonstration program on their own. Fortunately, these people tend to be highly mobile. They are accustomed to looking for the best places to conduct their research or find employment without regard to state borders. Their mobility is demonstrated in a special tabulation from the Survey of Doctorate Recipients that was prepared for this study (see Table 4-1): 70 percent of science and mathematics PhDs who are U.S. citizens or permanent residents leave the state where they earned their degree immediately after graduation. That is, of the 518,408 U.S. citizens or residents who received PhDs in 1957-1999, only 30 percent stayed in the state in which they received their degrees for their first employment. Only a national program can truly respond to the interests of prospective fellows and offer a

TABLE 4-1 Mobility of Mathematics and Science Doctorates After Graduation, 1957-1999 (U.S. Citizens and Permanent Residents)

State of PhD	Stayed in State (number)	Moved Out of State (number)	Total (number)	Stayed in State (percent)
Alabama	1,387	2,556	3,943	35.2
Alaska	98	129	227	43.2
Arizona	1,574	5,303	6,877	22.9
Arkansas	309	1,081	1,390	22.2
California	29,429	37,373	66,802	44.1
Colorado	3,139	7,129	10,268	30.6
Connecticut	1,682	7,098	8,780	19.2
Delaware	352	1,577	1,929	18.2
District of Columbia	1,875	5,260	7,135	26.3
Florida	3,451	8,630	12,081	28.6
Georgia	2,434	6,173	8,607	28.3
Hawaii	509	1,148	1,657	30.7
Idaho	210	698	908	23.1
Illinois	7,976	22,482	30,458	26.2
Indiana	2,559	14,000	16,559	15.5
Iowa	1,801	7,820	9,621	18.7

TABLE 4-1 Continued

State of PhD	Stayed in State (number)	Moved Out of State (number)	Total (number)	Stayed in State (percent)
Kansas	1,119	4,832	5,951	18.8
Kentucky	833	2,499	3,332	25.0
Louisiana	1,536	3,823	5,359	28.7
Maine	148	366	514	28.8
Maryland	3,623	8,288	11,911	30.4
Massachusetts	9,170	19,972	29,142	31.5
Michigan	5,339	16,031	21,370	25.0
Minnesota	2,656	7,062	9,718	27.3
Mississippi	705	2,182	2,887	24.4
Missouri	2,575	6,494	9,069	28.4
Montana	247	992	1,239	19.9
Nebraska	780	2,289	3,069	25.4
Nevada	145	386	531	27.3
New Hampshire	278	1,413	1,691	16.4
New Jersey	3,069	7,904	10,973	28.0
New Mexico	949	1,794	2,743	34.6
New York	19,888	34,016	53,904	36.9
North Carolina	4,157	10,439	14,596	28.5
North Dakota	241	927	1,168	20.6
Ohio	6,706	13,280	19,986	33.6
Oklahoma	1,443	4,164	5,607	25.7
Oregon	1,440	5,061	6,501	22.2
Pennsylvania	8,045	17,644	25,689	31.3
Rhode Island	628	2,999	3,627	17.3
South Carolina	895	2,522	3,417	26.2
South Dakota	101	450	551	18.3
Tennessee	2,207	5,446	7,653	28.8
Texas	9,634	16,233	25,867	37.2
Utah	1,472	4,205	5,677	25.9
Vermont	219	685	904	24.2
Virginia	2,473	6,967	9,440	26.2
Washington	2,763	7,598	10,361	26.7
West Virginia	330	1,251	1,581	20.9
Wisconsin	2,918	11,100	14,018	20.8
Wyoming	183	937	1,120	16.3
Total	157,700	360,708	518,408	30.4

SOURCE: Special tabulation from the Survey of Doctorate Recipients prepared for the committee by the Office of Scientific and Engineering Personnel, National Research Council.

range of choices to the institutions and schools that will be part of the demonstration program.

Program Duration

The proposed demonstration program would allow flexibility in the design of the local preparation programs, with the following constraint: that the institutions that choose to participate in the program will provide standards-based teacher education programs that integrate a variety of classroom teaching experiences with coursework. The coursework would include educational theory and the results of empirical research on how children learn science and mathematics. The courses would be structured as graduate-level seminars or other approaches that are appropriate for PhD-level teacher candidates. These programs are likely to be undertaken by institutions as add-ons to existing traditional or nontraditional teacher education programs, though some might design new programs for the fellows. Since one goal of the demonstration program is to provide the opportunity for state certification, the institutions will need to provide appropriate courses or other resources that allow for the fellows to become certified to teach in one or more states.

Each fellow will participate in a 2-year experience. One program might feature a full year of classroom-based study followed by a full year of school-based teaching, designed to lead to meeting qualifications for certification. Another program might decide its needs are better met by combining the classroom-based study and school-based teaching in the first year, and providing the fellows with opportunities to work in a wide variety of settings in the second year. Within the framework proposed by the committee, many other structures for the local programs would be possible.

The institutions that choose to participate in this demonstration program will likely already have ongoing partnerships with local schools, school districts, or larger geographical educational entities. If not, they will need to develop them. These partners will provide the settings for the school portions of the program, and they would be expected to assign an appropriate master teacher to act as a mentor to each fellow.

Teacher Preparation and Certification

We have already discussed the kinds of knowledge that new and recent PhDs in science, mathematics, and engineering need to master to become effective K-12 teachers, and Appendix C describes some nontraditional teacher preparation programs. A more extensive source of information on the preparation required for effective teaching is a recent report of the NRC (2000b), which reviewed the research results and recommendations of professional organizations.

As described in Chapter 3, as an addition to the deep content knowledge in their fields, the fellows need to develop both pedagogical knowledge and pedagogical content knowledge: that is, how to teach in general, and how to teach particular subjects. Thus, for example, they need to learn how to teach students of different ages, with different prior knowledge, and from different cultural and socioeconomic backgrounds. They also need to learn about classroom management and the structure of K-12 education, with its national, state, and local dimensions. In the committee's proposed demonstration program, this pedagogical knowledge would be gained in both classroom-based and school-based learning and teaching experiences.

One question with which the committee wrestled is certification: What role, if any, should certification have in the program? There are two issues involved in this question: one concerns the locus for certification; the other concerns the possibility of careers other than teaching in K-12 education.

On the first issue, some participants at the committee's workshop suggested that the demonstration program might aim at setting certification standards at a national level. However, teacher certification has always been the purview of the states, and there was considerable doubt expressed by other participants that states would—or could, legally—let an outside body set certification requirements. Moreover, the requirements for certification are variable across the nation (see Feistritzer and Chester, 2000). Yet it is critically important that the fellows obtain certification if they are to teach in public schools. The committee believes that the appropriate way to deal with certification in the demonstration program is to require that the 2-year program include all of the courses and other activities that are necessary for fellows to obtain state certification, at least in the state in which their preparation takes place.

On the second issue, some workshop participants questioned whether the demonstration program should require the courses and activities for certification, since not all positions in K-12 education require teaching certification. However, the committee believes that if the fellows are to become leaders in improving K-12 science, mathematics, and technology education—whether directly in the classroom or in some other capacity—

the certification experience will provide credibility for them to work effectively with teachers. The committee therefore strongly endorses the position that all fellows should achieve certification, and as noted above, that the institutions that prepare them should ensure that the fellows can meet certification requirements. However, actually obtaining certification should not be a requirement for remaining in the demonstration program or successfully "graduating." The committee notes that experience with scientists and engineers from defense-related companies who prepared to become teachers under the Defense Reinvestment Initiative (see NRC, 1999a) shows that fellows who do not achieve certification during their 2 years in a program may become certified after the program is over.

Support for the Fellows

To make the demonstration program attractive to prospective participants, the committee proposes that it be structured as a 2-year postdoctoral fellowship. By the time they have completed their work, PhDs have invested a significant amount of their time and funds in their professional preparation, and virtually all of the options open to them involve being paid or financially supported in some way. It is not realistic to ask them to then undertake two additional years of preparation on their own for one of the options—to help improve the nation's K-12 education—that may be as important to the nation as it is to the fellows. Moreover, the fellows will be providing significant labor to their home school district, particularly in the second year of the program, so in that year the stipend can be viewed as in large part a salary.

The committee believes that the first-year stipend should be awarded as a prestigious national fellowship. The second-year support might be in the form of a position in a school district as a teacher or in some other supporting role in K-12 education, funded at least in part by the school, district, or other education institution that is "employing" the fellow. Having the institution contribute some or all of the support for the second-year stipend would also demonstrate their commitment to the project.

We estimate that the fellowship will cost a total of about $56,000 per year per fellow, which will cover the recommended stipend of $35,000, plus health and related employee benefits, and travel funds. Travel funds are important so that the fellows can attend conferences and other gatherings to keep in touch with both their scientific and their teaching colleagues throughout the country.

To estimate the cost, the committee first examined how stipend levels are set for postdoctoral awards in the AAAS Congressional Fellowship Program of the American Association for the Advance of Science (AAAS) and the Associateship Program of the National Research Council (NRC). In the AAAS program, the stipends are commensurate with the salaries for entry-level employees of the host institution who have similar job descriptions. For the NRC associateships, the host laboratories set the stipend levels; the program suggests that they use the federal government's grade 11 step-1 salary level as a benchmark for setting the stipend levels, but they are free to set the level above or below that target. Based on the review of teacher salaries in New Jersey, North Carolina, Texas, California, and Washington presented in the Phase I report (NRC, 2000a), we estimate that the average annual salaries in those states for entry-level teachers at the secondary level who hold PhDs is about $35,000. A stipend at this level is close to the $37,128 per year salary that graduate students and postdoctoral fellows surveyed for the report would expect to receive as a secondary school teacher. Therefore, the stipend level suggested is not expected to be a major barrier in recruitment, especially in states with pay scales similar to those of the states reviewed in Phase I. To the base $35,000 per year stipend the committee adds an estimated $15,000 for benefits, $3,000 to cover relocation expenses, and $3,000 to cover travel, for an average total annual cost of about $56,000 per PhD. However, because the cost of living and pay scale for teachers and other educators is not uniform across the nation, a geographic variation in the level of the stipends might be used for the demonstration program.

In addition to financial support, the support of mentors—most of whom will be master teachers—will be critical to the program. As discussed in Chapter 3, teaching is a challenging and demanding profession, and classroom experience is an indispensable part of a teacher's learning. Some of that experience and learning can best be passed on from master teachers to novice teachers, which both makes learning proceed more rapidly and gives the new teachers more confidence in their ability to become high-quality teachers.

OTHER PROGRAM CHARACTERISTICS

In addition to the key elements discussed above, several other features will need to be carefully designed in the program: recruitment, selection and placement, and teacher preparation.

Recruitment

An important first step in the proposed program is identifying and recruiting the target population, new and recent PhDs in science, mathematics, and engineering. Many postdoctoral fellowship programs define "recent" to be within the last 5 years, and the committee believes that that time frame is appropriate for this program. Furthermore, the committee believes that doctoral students should be allowed to apply to the program if they have not completed their PhD at the time of application but will do so by the beginning of the program. That way, the PhDs can go immediately from their graduate studies to the fellowship program.

The group that discussed recruitment at the committee's public workshop believed the population of recent PhDs in science, mathematics, and engineering is reasonably well defined, so that recruitment efforts can be targeted to the appropriate trade journals and organizations. Word of mouth is likely to be an important part of recruitment, particularly after the program begins, when current and former participants will be excellent resources for recruitment.

Another subject that arose in the workshop discussion concerned possible tension between: (1) an intent of the program to address the shortage of highly qualified lead teachers and (2) the desire of some professionals in the field to strengthen the infrastructure of K-12 education in science, mathematics, and technology in other ways. This tension might be reflected in the recruitment process to the extent that some applicants might be more interested in other positions in K-12 education, such as resource specialists, and might not remain in teaching positions after the 2-year fellowship. The group's discussion centered on PhDs who are particularly interested in teaching, but it concluded that other careers in K-12 education should not be excluded from consideration.

National recruitment efforts might include program announcements through a variety of Internet-based listservs, a specialized website, an informational brochure, and advertisements in *Science*,[1] *Nature, Chronicle of*

---

[1]There is some anecdotal evidence to suggest the journal *Science* might be an effective recruitment medium. An alternative certification program for PhDs developed as a partnership between the National Institutes of Health (NIH) and the Montgomery County [Maryland] Public Schools (see Appendix C) was described very briefly in the letters section of the October 13, 2000, issue of the magazine. Although the program is intended only for PhDs who are already at NIH's Bethesda campus, the letter generated a national response: 30 people made casual inquiries and 13 others sent in material that they hoped would give them entry into the program.

Higher Education, and discipline-based journals. Scientific, mathematical, and engineering professional societies might also play a role in recruitment. One model might be the Science and Technology Policy Fellowships Program of the American Association for the Advancement of Science (AAAS, 2002).[2] In that program, AAAS recruits, selects, and funds two of its own fellows and runs an umbrella program for the fellows recruited, selected, and funded by about 30 other national scientific and engineering societies.

Special recruiting efforts might be needed to ensure that the program is known to all groups. For example, participants at the committee's workshop noted that there are mentoring networks in some minority communities that might be very valuable in recruiting for this program. It is most important, of course, that graduating PhDs be aware of the program and that it is well advertised within graduate school academic communities.

Selection and Placement

In addition to the criterion that prospective fellows must be new or recent PhDs in science, mathematics, or engineering, the program should, to the extent possible, identify PhDs who have a strong desire to have a career in K-12 education. Toward that goal, applicants should provide evidence of an interest in K-12 education. Evidence might include prior participation in formal K-12 activities (e.g., classroom teaching, classroom observation, tutoring, literacy or partnership programs, etc.) or informal K-12 activities (e.g., programs in aquaria, zoos, museums, environmental centers, etc.). Furthermore, given the particular shortage of highly qualified teachers in urban and rural school districts, applicants should demonstrate interest and desire in teaching *all* students—from a variety of socioeconomic and cultural backgrounds and from different levels of academic preparation and achievement.

Although it is important to select fellows who have a demonstrated commitment to K-12 education, it must be kept in mind that PhD students often carry heavy burdens to complete their work, so that they may not have had significant amounts of time for other activities. The people

---

[2]This program is open to people with PhDs or an equivalent degrees in the social, physical, or biological sciences, or who are engineers with a master's degree and 3 years of post-degree professional experience. The 1-year fellowships, based in Washington, DC, are designed to provide participants with public policy learning experience and to bring technical backgrounds and external perspectives to decision making in the U.S. government.

involved in the selection process will have to develop realistic and flexible requirements for the criterion of a commitment to K-12 education. Applicants might be asked to submit letters of recommendation and an essay that describes their ideas about teaching, their ideas for integrating scientific experiences into classroom teaching, and long-term career goals. The applicants might also be required to take the Gallup Teacher Perceiver (see Appendix E) or similar instrument to assess their interests and skills for classroom teaching. Selection should be through a peer review process: the evaluators should include master K-12 teachers, distinguished scientists, mathematicians, and engineers, and distinguished teacher educators.

The selection process will depend on the detailed structure of the demonstration program. There are three basic selection and placement scenarios that the committee believes should be considered. In the first scenario, the institution of higher education that will provide the first year's preparation and the local schools or schools districts would be responsible for selecting the fellows. This option may be attractive to the institutions and schools because it gives them complete control over which PhDs are accepted into their programs. But it has two potentially serious negative consequences: the selection criteria might vary significantly from site to site, and PhDs might find it difficult to know about or apply to what are in effect many different programs. In a second scenario, at the other end of a spectrum, the selection and placement of the fellows would be handled entirely by the national-level entity that administers the program. Such an arrangement might provide economies of scale, promote the use of uniform selection criteria, and allow applicants to apply for several different institutions or places. But it, too, has several serious negative consequences: the institutions and schools that are the key places for the preparation might well not be willing to give up their ability to select candidates for their own programs. The third scenario is a joint process that involves both a national entity and the local providers. In this case, the precise roles for each partner would need to be worked out at the beginning in the detailed design phase.

One model for a joint selection and placement process is the National Resident Matching Program (NRMP) for medical training. The applicants to graduate medical programs and the programs submit ranked lists of either their preferred programs (from applicants) or their preferred applicants (from programs) to the NRMP organization. When all of the rank lists are received by the NRMP, it compares the lists and undertakes a process to find the best matches (see Appendix F for a description of the matching process).

A second model is the selection and placement process used by the NRC Associateship Programs (see NRC, 2001b). One component of that program, for scientists and engineers who are less than 5 years past receiving their PhDs, places associates in 1- to 3-year research postdoctoral positions at federal facilities across the nation. The host laboratory sets the length of the postdoctoral appointment. Candidates for the associateship submit with their application materials a proposal for research with a specific adviser at one of the participating federal laboratories. Applicants may submit up to three applications for each review (there are three reviews per year). The proposal is first screened by the potential adviser, who indicates whether the candidate should be considered by the NRC in its formal review process. The NRC then convenes peer-review panels to rank order the research applications that have passed the screening by the advisers. The associateships are awarded to the applicants according to the rank order of their application compared with others for the same federal laboratory.

A third possible model is the selection and placement process used by the nine AAAS Science and Technology Policy Fellowships Programs. Although the selection process varies somewhat among programs, they all screen the applications through an informal peer-review mechanism involving former fellows. The peer-review process leads to an initial cut in the applicant pool. Selection committees of relevant experts who are familiar with the program and with science and technology policy then meet to determine which applicants are invited to Washington, DC, for interviews. In most programs, these committees meet again to determine which candidates are eligible to become fellows. The placement is decided through interviews between the fellows thus selected and the offices that have expressed an interest in hosting a fellow, and the AAAS assists the fellows in finding a suitable placement.

Mentoring and Leadership Preparation

If the fellows are to be effective teachers and leaders in K-12 education, they will need to be knowledgeable about and comfortable with this system's customs and values. A key to this acculturation process—as well as a valuable teaching method—is through mentoring. This type of support can provide experience, advice, and feedback on everything from lesson plans and accommodating parents to understanding and working with boards of education or museum docents.

Participants at the committee's workshop identified three general types of mentoring that they suggested would be valuable to fellows:

- transitional mentoring, involving general issues common to all fellows in the program (e.g., understanding different value systems, acquiring pedagogical skills);
- navigational mentoring, involving day-to-day issues that are relevant to the fellows' local environment (e.g., accessing local resources, dealing with the parent organization); and
- developmental mentoring, involving individual professional development (e.g., building a network of colleagues, developing leadership and classroom management skills).

The participants were assuming that the demonstration program would have both national and local components and suggested that all partners have a role to play in providing mentoring. In particular, they thought that the national program should play the major role in providing transitional mentoring, the school district should provide navigational mentoring, and that other partners (e.g., universities, museums) should take the lead in providing developmental mentoring. They also indicated that it is most important for every fellow to have a mentor in the form of a master teacher who is on site every day and would be involved in all three levels of mentoring.

The committee believes that the fellows should be expected to contribute to K-12 educational scholarship[3] in ways that demonstrate a new kind of leadership in the classroom and help build the standards-based infrastructure. There are two distinct concepts denoted by the word "leadership." In its hierarchical form, leadership is conferred upon people who occupy particular positions in an organization, such as school principals or university deans. Although it is anticipated that some of the fellows may ultimately occupy a variety of positions in the K-12 educational system, it is likely that others will remain in the classroom for their entire careers. The other form of leadership refers to the actions of a group of people working together to bring about change in a system toward common goals and

---

[3] The PhDs might conduct research in the areas identified in Chapter 3 as being critical to their success as K-12 educators. The topics are pedagogical content knowledge, pedagogy, context and diversity, classroom management, and schooling.

values (see, e.g., Astin and Astin, 2000). This is the type of leadership preparation that the committee believes should be part of the demonstration program—building on the unique skills that PhDs possess so that they may become effective leaders in improving standards-based K-12 science, mathematics, and technology education for all students. It would not be targeted at developing future school principals, school district superintendents, or other "leaders."

STRUCTURE

Although the Phase I report of this project (NRC, 2000a) raised the possibility of state programs, the committee concludes that the proposed demonstration program should be a national one. As discussed in Chapters 2 and 3, both sides of the equation for which this program is designed—the needs in K-12 science, mathematics, and technology education, and the population of new and recent PhDs in science, mathematics, and engineering—are national in scope. In order to offer the most opportunities to both fellows and schools, a national program is needed. The committee also reiterates its belief that the problem deserves national attention.

Other reasons for a national program concern economies of scale. Recruitment, including advertising, selection, and placement are all functions that can be carried out more efficiently on a national basis. Applicants can apply to several different places for their preparation; institutions of higher education and schools can choose from a national pool. Having a national program also means that there will be enough people in the same program to make it possible to design an evaluation that can yield clear findings.

The committee envisions a program with 30-60 fellowships available each year to sequential overlapping cohorts of PhDs for the duration of the demonstration program. The committee believes that 10 years is a reasonable length of time for this demonstration program—to determine if it really works, and at what cost. The committee is aware that it may be necessary to precede the demonstration program with a smaller initial program—say, 10-15 fellows for four years—to show that there is sufficient interest among PhDs and in the K-12 community to warrant the larger effort.

There will have to be some kind of national structure to administer the program. That national structure may also serve as a convening mechanism for the fellows—perhaps holding yearly or more frequent meetings so the fellows can share experiences and learn from each other, and engage in leadership activities. It may also be useful to have yearly meetings with others involved in the program, for example, the mentors. The detailed structure of the proposed program will have to be developed in the next phase of this project, with appropriate input from all the communities that will be involved in the effort: potential fellows, institutions of higher education, schools, and master teachers.

FUNDING

The major program expense will be the $56,000 per year that the committee estimates to be required to support each fellow (see above). To estimate the administrative costs of the program, the committee looked again at the AAAS fellowship programs, particularly the Congressional Fellowship Program, and at the NRC Associateship Program. In 2001, the overall operating costs for the AAAS program were about $2,900 per fellow plus $1,000-$3,000 per fellow for the selection process; in the NRC program, the costs for both program administration and selection were about $9,000 per associate (about 15-17 percent of the total costs). Neither program includes the type of national meeting that the committee believes should be considered for this program, so the overall costs are likely to be somewhat higher than the $5,000-$9,000 per person for the AAAS and NRC programs. The committee offers the figure of $11,000 as a rough estimate; a more refined estimate will need to be developed in the detailed program design.

Assuming that the first-year stipend and all administrative costs would be supported at the national level, the cost for each fellow would be $67,000 for the first year (covering all costs) and $14,000 for the second year (travel and administrative costs). The committee believes that the school, school district, or state would cover the $35,000 stipend and $15,000 in benefits for the fellows in their second year, when they will be working in local settings. Thus, for a yearly cohort of 30 fellows the cost of the first year of the program would be a little more than $2 million; for a cohort of 60 fellows, a little more than $4 million. For the second and subsequent years, the national annual cost would be about $2,425,000 for 30 fellows per cohort and about $4,850,000 for 60 fellows per cohort.

The committee expects that the institutions of higher education and local schools and school districts will incur some costs related to designing and conducting their parts of the fellowship program. However, it is difficult to estimate those costs because of the variety of ways those institutions might design programs and because they may have similar programs already in place, which would reduce the costs of adding this program. The committee is assuming that these partners in the demonstration program would contribute the necessary funding for their parts.

The committee also has not included in its estimate the costs of program evaluation and administrative overhead. Although these costs could be significant, their magnitudes will depend upon the details of how the demonstration program is implemented in Phase III.

At the committee's workshop, participants suggested that support for the demonstration program might come from a mix of federal, state, school district, private foundation, and industry sources. For example, foundations might provide seed money or start-up money, while, as just noted, states or school districts would be expected to support some local components of the program. One possible source of funding for the fellows' stipends might come through a sponsorship by professional societies, as is the case with the AAAS Science and Technology Policy Fellowship Programs.

Because the proposed program is a national one, federal support would likely be needed, and it would be appropriate, reflecting the national scope of the issues. Looking at similar programs supported by the National Science Foundation (NSF), that agency might be a source of support for the demonstration program. Other federal agencies that have an important interest in the issues being addressed are the National Institutes of Health (NIH) and the Department of Education. The three agencies, and perhaps others, might work in collaboration on this program, which cuts across areas of interest to all of them.

EVALUATION

Central to any demonstration program is its evaluation. The findings of studies of program evaluation make clear that an evaluation plan has to be developed and implemented with the implementation of a demonstration program itself: it cannot be tacked on later. An evaluation plan must also reflect the actual implementation of the program: it cannot reflect only the plan and design of the program. Thus, the evaluation plan for the committee's proposed demonstration program must be designed as the details of the program are designed and then refined as the program is implemented.

The committee notes that several groups should be involved in the development of the evaluation plan to ensure that it is appropriate for the program and can be done. They include the institutions of higher education that will provide some preparation to the PhDs and the local schools or school districts that are the setting for the teaching parts of the program. Most important, an organization that has experience and proven competence in designing evaluation plans for K-12 educational programs should be involved.

NEXT STEPS

This report marks the conclusion of the second phase of a project to explore the possibilities and design of a demonstration program to attract PhDs in science, mathematics, and engineering to careers in K-12 education. The twin goals of this project are to offer a challenging career opportunity to new and recent PhDs and to tap their unique talents to improve the nation's educational system. The first phase showed that significant numbers of PhDs would be interested in careers in K-12 education. This report considers how to simultaneously meet the needs and interests of the PhDs and the nation's schools, and it specifies some of the design criteria for the demonstration program.

The program the committee proposes—a national 2-year fellowship for new and recent PhDs—if carefully implemented and rigorously evaluated, will provide clear evidence of its benefits and costs and can, if the benefits outweigh the costs, guide the development of large-scale, long-term efforts in this area. It is clear that just having the distinguished fellowship is not sufficient. Careful implementation will involve strong connections with participating local schools and districts, as well as strong and appropriate programs at the participating institutions of higher education and continuing contacts among the fellows. The proposed program carries the expectation that a new community of educators will develop among the fellows, who carry the cultures of both K-12 education and their disciplines, and so contribute to a high level of science, mathematics, and technology in K-12 education across the nation. The committee strongly endorses moving to Phase III of the project: calling together potential funders and program designers to implement the program.

References

Alberts, B. (2000). Some thoughts of a scientist on inquiry. In Minstrell, J. and E. H. van Zee (Eds.). *Inquiring into inquiry learning and teaching in science* (pp. 3-13). Washington, DC: American Association for the Advancement of Science.

American Association for the Advancement of Science. (1990). *Science for all Americans.* New York: Oxford University Press.

American Association for the Advancement of Science. (1993). *Benchmarks for science literacy.* New York: Oxford University Press.

American Association for the Advancement of Science. (2002). *AAAS Science and Technology Policy Fellowships.* Available: http://fellowships.aaas.org/ [May 13, 2002].

American Chemical Society. (2001). *ChemCom® chemistry in the community.* Available: http://198.110.10.57/ChemCom/ [September 26, 2001].

American Institute of Physics. (1999). *Maintaining momentum.* College Park, MD: Author. Available: http://www.aip.org/statistics/trends/reports/hsreport.pdf [October 25, 2001].

Association of Science-Technology Centers, Incorporated. (2001). *ASTC sourcebook of science center statistics.* Washington, D.C.: Author.

Astin, A.W., and Astin, H.S. (2000). *Engaging higher education in social change.* Battle Creek, MI: W.K. Kellogg Foundation.

Clewell, B., and Forcier, L. (2000). *Increasing the number of mathematics and science teachers: A review of teacher recruitment programs.* Paper commissioned by the National Commission on Mathematics and Science Teaching for the 21st Century. Available: http://www.ed.gov/americacounts/glenn/ClewellForcier.pdf [September 26, 2001].

Cohen, K.C. (Ed.). (1997). *Internet links for science education: Student-scientist partnerships.* New York, NY: Plenum Press.

Darling-Hammond, L. (1999). *Teacher quality and student achievement: A review of state policy evidence.* New York: Center for the Study of Teaching and Policy. Available: http://depts.washington.edu/ctpmail/Publications/PDF_versions/LDH_1999.pdf [October 1, 2001].

Eisenhower National Clearinghouse. (2001a). *State frameworks.* Available: http://www.enc.org/professional/standards/state/ [September 25, 2001].

Eisenhower National Clearinghouse. (2001b). *Welcome to ENC!* Available: http://www.enc.org/ [September 26, 2001].

Feistritzer, C.E., and Chester, D.T. (2000). *Alternative teacher certification: A state-by-state analysis.* Washington DC: National Center for Education Information.

Fiske, P.S. (2001). *Put your science to work: The take-charge career guide for scientists.* Washington, D.C.: American Geophysical Union.

Gallup Organization. (2002). *Perceiver interviews* Available:. HtmlResAnchor http://education.gallup.com/select/perceiverInterviews.asp [January 31, 2002].

Goodnough, A. (2001). *'S' is for satisfactory, not for satisfied, on teacher's sentimental journey.* New York Times, July 1. Page 1.

Gregorian, V. (2001). *Teacher education must become colleges' central preoccupation.* The Chronicle of Higher Education, August 17, B7.

Hare, D., and Heap, J.L. (2001). *Effective teacher recruitment and retention strategies in the midwest: Who is making use of them?* North Central Regional Educational Laboratory. Available: http://www.ncrel.org/policy/pubs/html/strategy/section1.htm [February 1, 2002].

Haury, D.L. (1998). Preparing scientists and mathematicians for a second career in teaching. In Risacher, B.F. (Ed.). *Scientists and mathematicians become school teachers.* Columbus, OH: Eric Clearinghouse for Science, Mathematics and Environmental Education.

Howard Hughes Medical Institute. 2001. *Biointeractive.* Available: http://www.biointeractive.org/ [September 26, 2001].

Hurd, P. DeH. (1997). *Inventing science education for the new millennium.* New York, NY: Teachers College Press

Ingersoll, R. (1999). The problem of underqualified teachers in American secondary schools. *Educational Researcher, 28*,26-36.

International Technology Association. (2000). *Standards for technological literacy.* Reston, VA: Author.

Krajcik, J., Czerniak, C., and Berger, C. (1999). *Teaching children science: A project-based approach.* New York: McGraw-Hill College.

Kreeger, K.Y. (1998). *Nontraditional careers in science.* Philadelphia, PA: Taylor & Francis.

Minstrell, J. (Ed.). (2000). *Inquiring into inquiry learning and teaching in science.* Washington, DC: American Association for the Advancement of Science.

Munn, M., Skinner, P.O., Conn, L., Horsma, H.G., and Gregory, P. (1999). The involvement of genome researchers in high school. *Genome Research, 9,* 597-607.

National Aeronautics and Space Administration. (2001). *Education programs.* Available: http://education.nasa.gov/ [October 1, 2001].

National Assessment Governing Board. (2001). *Spotlight on the nation.* Available: http://nagb.org/pubs/students/nation.html [September 28, 2001].

National Association of State Boards of Education. (1998). *The numbers game: Ensuring quantity and quality in the teaching work force.* Alexandria, VA: Author.

National Center for Education Statistics, U.S. Department of Education. (1999a). *Highlights from TIMSS.* NCES 1999-081. Washington, DC: U.S. Government Printing Office.

REFERENCES

National Center for Education Statistics, U.S. Department of Education. (1999b). *Predicting the need for newly hired teachers in the United States to 2008–09*. NCES 1999-026. Washington, DC: U.S. Government Printing Office. Available: http://nces.ed.gov/pubs99/1999026.pdf. [October 1, 2001].

National Center for Education Statistics, U.S. Department of Education. (2000). Pursuing excellence: Comparisons of international eighth-grade mathematics and science achievement from a U.S. perspective, 1995 and 1999. P. Gonzales, C. Calsyn, L. Jocelyn, K. Mak, D. Kastberg, S. Arafeh, T. Williams, and W. Tsen. NCES 2001-028. Washington, DC: U.S. Government Printing Office.

National Center for Teaching and America's Future. (1996). *Doing what matters most: Teaching for America's future*. New York: Author.

National Commission on Mathematics and Science Teaching for the 21st Century. (2000). *Before it's too late*. Jessup, MD: Education Publications Center.

National Council of Teachers of Mathematics. (1989). *Principles and standards for school mathematics*. Reston, VA: Author.

National Council of Teachers of Mathematics. (2000). *Principles and standards for school mathematics*. Reston, VA: Author.

National Institutes of Health. (2001). *NIEHS extramural K-12 education programs*. Available: http://www.niehs.nih.gov/od/k-12/allextra.htm [October 1, 2001].

National Research Council. (1995). *Reshaping the graduate education of scientists and engineers*. Committee on Science, Engineering and Public Policy. Washington, DC: National Academy Press. Available: HtmlResAnchor http://www.nap.edu/catalog/4935.html. [February 21, 2002].

National Research Council. (1996a). *Careers in science and engineering: A student planning guide to grad school and beyond*. HtmlResAnchor Committee on Science, Engineering, and Public Policy. Washington, DC: National Academy Press. Available: http://www.nap.edu/catalog/5129.html. [February 21, 2002].

National Research Council. (1996b). *National science education standards*. National Committee on Science Education Standards and Assessment. Washington, DC: National Academy Press. Available: http://www.nap.edu/catalog/4962.html. [February 21, 2002].

National Research Council. (1996c). *The role of scientists in the professional development of science teachers*. Committee on Biology Teacher Inservice Programs. Washington, DC: National Academy Press. Available: http://www.nap.edu/catalog/2310.html. [February 21, 2002].

National Research Council. (1998). *Trends in early careers of life scientists*. Committee on Dimensions, Causes, and Implications of Recent Trends in the Careers of Life Scientists. Washington, D.C.: National Academy Press. Available: http://www.nap.edu/catalog/6244.html. [February 21, 2002].

National Research Council. (1999a). *Final report to the U.S. Department of Defense on the defense reinvestment initiative*. Defense Reinvestment Initiative Advisory Board. Washington, DC: National Academy Press. Available: http://www.nap.edu/catalog/9691.html [October 28, 2001].

National Research Council. (1999b). *How people learn: Brain, mind, experience, and school.* Bransford, John D., Brown, Ann L., and Cocking, Rodney R. (Eds.). Committee on Developments in the Science of Learning. Washington, DC: National Academy Press. Available: http://books.nap.edu/catalog/6160.html. [February 21, 2002].

National Research Council. (1999c). *How people learn: Bridging research and practice.* Donovan, M.S., Bransford, J.D., and Pellegrino, J.W. (Eds.). Committee on Learning Research and Educational Practice. Washington, DC: National Academy Press. Available: http://books.nap.edu/catalog/9457.html. [February 21, 2002].

National Research Council. (1999d). *Improving student learning: A strategic plan for education research and its utilization.* Committee on a Feasibility Study for a Strategic Education Research Program. Washington, DC: National Academy Press. Available: http://www.nap.edu/catalog/6488.html. [February 21, 2002].

National Research Council. (2000a). *Attracting science and mathematics Ph.D.s to secondary school education.* Committee on Attracting Science and Mathematics Ph.D.s to Secondary School Teaching. Office of Scientific and Engineering Personnel. Washington, DC: National Academy Press. Available: http://www.nap.edu/catalog/9955.html. [February 21, 2002].

National Research Council. (2000b). *Educating teachers of science, mathematics, and technology: New practices for a new millennium.* Committee on Science and Mathematics Teacher Preparation. Washington, DC: National Academy Press. Available: http://www.nap.edu/catalog/9832.html. [February 21, 2002].

National Research Council. (2000c). *Enhancing the postdoctoral experience: A guide for postdoctoral scholars, advisers, institutions, funding organizations, and disciplinary societies.* National Academy of Sciences, National Academy of Engineering, Institute of Medicine. Washington, DC: National Academy Press. Available: http://www.nap.edu/catalog/9831.html. [February 21, 2002].

National Research Council. (2000d). *Inquiry and the national science education standards: A guide for teaching and learning.* Steve Olson and Susan Loucks-Horsley, (Eds). Committee on the Development of an Addendum to the National Science Education Standards on Scientific Inquiry. Washington, DC: National Academy Press. Available: http://www.nap.edu/catalog/9596.html. [February 21, 2002].

National Research Council. (2001a). *Classroom assessment and the national science education standards.* J. Myron Atkin, Paul Black, Janet Coffey, (Eds.). Committee on Classroom Assessment and the *National Science Education Standards.* Washington, DC: National Academy Press. Available: http://www.nap.edu/catalog/9847.html. [February 21, 2002].

National Research Council. (2001b). *NRC postdoctoral associateships.* Available: http://www4.nationalacademies.org/pga/rap.nsf [October 29, 2001].

National Resident Matching Program. (2002). *National resident matching program.* Available: http://www.nrmp.org. [January 31, 2002].

National Science Foundation. (2000). *Centers for learning and teaching (CLT), program solicitation, NSF 00-148.* Available: http://www.nsf.gov/pubs/2000/nsf00148/nsf00148.htm [September 25, 2001].

National Science Foundation. (2001a). *NSF postdoctoral fellowships in science, mathematics, engineering, and technology education (PFSMETE).* Available: http://www.ehr.nsf.gov/dge/programs/pfsmete/ [September 25, 2001].

REFERENCES

National Science Foundation. (2001b). *NSF graduate teaching fellows in K-12 education (GK-12)*. Available: http://www.ehr.nsf.gov/dge/programs/gk12/ [September 25, 2001].

National Science Foundation. (2001c). *Centers for learning and teaching (CLT)*. Available: http://www.interact.nsf.gov/cise/descriptions.nsf/Pages/EC85AE4B5110C7BF85256A02 00659590 [September 25, 2001].

National Science Foundation. (2001d). *Advanced technological education (ATE)*. Available: http://www.ehr.nsf.gov/ehr/due/programs/ate [September 26, 2001].

National Science Foundation. (2001e). *Implementation sites*. Available: http://www.ehr.nsf.gov/esie/resources/impsites.asp [September 26, 2001].

National Science Foundation. (2001f). *Interagency education research initiative (IERI) program solicitation*. Available: http://www.nsf.gov/pubs/2000/nsf0074/nsf0074.htm [October 1, 2001].

National Science Resources Center, National Academy of Sciences, and Smithsonian Institution. (1997). *Science for all children: A guide to improving science education in your school district*. Center for Science, Mathematics, and Engineering Education. Washington, DC: National Academy Press. Available: http://www.nap.edu/catalog/4964.html [October 1, 2001].

Nyquist, J. (1999). *Re-envisioning the Ph.D. panel: National surveys and studies*. Seattle, WA: University of Washington.

Nyquist, J., and Woodford, B. (2000). *Re-envisioning the Ph.D.: What concerns do we have?* Seattle, WA: University of Washington.

Raymond, M., Fletcher, S.F., Luque, J., (2001). *Teach For America: An Evaluation of Teacher Differences and Student Outcomes in Houston, Texas*. Available: http://www.teachforamerica.org/tfa/about/studies.html#teacher [February 28, 2002].

Reese, C.M., Miller, K.E., Mazzeo, J., and Dossey, J.A. (1997). *NAEP 1996 mathematics report card for the nation and the states*. Washington, DC: National Center for Education Statistics.

Robbins-Roth, C. (Ed.). (1998). *Alternative careers in science*. San Diego, CA: Academic Press.

Rosen, S., and Paul, C. (1997). *Career renewal*. San Diego, CA: Academic Press.

Sussman, A. (Ed.). (1993). *Science education partnerships: Manual for scientists and K-12 teachers*. San Francisco: University of California.

Tobias, S., Chubin, D., and Aylesworth, K. (1995). *Rethinking science as a career: Perceptions and realities in the physical sciences*. Tucson, AZ: Research Corporation.

U.S. Department of Agriculture. (2001). *Agriculture in the classroom*. Available: http://www.agclassroom.org/ [October 1, 2001].

U.S. Department of Education. (2000). *Our national education goals*. Available: http://www.ed.gov/Welcome/natgoals.html [September 28, 2001].

U.S. Department of Education. (2001). Office of Educational Research and Improvement. National Center for Education Statistics. *The nation's report card: Mathematics 2000*, NCES 2001–517, by J.S. Braswell, A.D. Lutkus, W.S. Grigg, S.L. Santapau, B. Tay-Lim, and M. Johnson. Jessup, MD: Education Publications Center. Available: http://nces.ed.gov/nationsreportcard/pdf/main2000/2001517.pdf [September 28, 2001].

U.S. Department of Education. (2002). *The nation's report card: Science highlights 2000.* NCES 2000-452. Jesup, MD: Education Publications Center. Available http://nces.ed.gov/nationsreportcard/pdf/main2000/2002452/df [May 13, 2002].

U.S. Department of Energy. (2001). *Office of science education.* Available: http://www.scied.science.doe.gov/scied/sci_ed.htm [October 1, 2001].

Appendices

Appendix A
Executive Summary*
Attracting Science and Mathematics PhDs to Secondary School Education

The United States is at a critical juncture in science and mathematics education. The U.S. Department of Education has projected that the nation's school systems will need to hire more than two million new teachers during the next decade. Finding qualified teachers of science and mathematics will pose a special challenge, as many school districts already find it difficult to recruit science teachers. This report examines whether recent Ph.D.s in science and mathematics might provide an additional resource for helping to meet the nation's need for qualified secondary school science and mathematics teachers in the coming years, while creating valuable connections between U.S. schools and our vibrant science and engineering communities.

BACKGROUND

The National Research Council (NRC) has been deeply involved in the last decade in efforts to improve the science and mathematics education of our nation's schoolchildren. The 1996 *National Science Education Standards* urged "changes in what students are taught, in how their performance is assessed, in how teachers are educated and keep pace, and in the relationship between schools and the rest of the community—including the nation's

---
*Can be found at http://www.nap.edu/catalog/9955.html.

scientists and engineers." It also emphasized the importance of "a new way of teaching and learning about science that reflects how science is done, emphasizing inquiry as a way of achieving knowledge and understanding about the world."

The NRC has followed the publication of the *Standards* with additional studies and programs that further explore key aspects of their implementation. In early 1999 the NRC launched a three-phase project to explore the feasibility of attracting scientifically trained Ph.D.s to positions in secondary school education as a possible mechanism to help improve science and mathematics education.

In launching this project the NRC assumed that Ph.D. training with its strong emphasis on experimental evaluation, quantitative approaches and mathematical content could potentially make a meaningful contribution to the implementation of an inquiry-based learning environment. Also, due to the large number of Ph.D.s who have experienced difficulty recently moving out of postdoctoral positions—especially in the life sciences—the nation has an unusual opportunity to attract these Ph.D.s to America's secondary school classrooms. Most Ph.D.s are well suited to the research careers they have chosen and should continue to pursue them. Yet there are many Ph.D.s whose training, personalities, and outlook would make them ideal candidates for secondary school teaching positions.

There are, of course, a number of potential obstacles to Ph.D.s taking secondary school teaching positions. These include the willingness of Ph.D.s to take education courses and obtain certification; the attitudes of professors, colleagues, mentors, high school principals, and other secondary school teachers; the potential opposition of teachers' unions; salary levels, and others. The purpose of the first phase of the project was to evaluate the possibilities and obstacles and to recommend possible incentives to overcome them.

SCOPE OF STUDY

The NRC project to attract science and mathematics Ph.D.s to secondary education was organized into three phases. The first phase was designed to tell us whether there was any interest among Ph.D.s in becoming secondary school teachers and, if so, what incentives states and school districts might use to attract them. This report summarizes the findings from these investigations, with suggestions to the committee overseeing phase two.

Phase two will use information gained from phase one and other information sources to help design state-based demonstration programs to attract science and mathematics Ph.D.s to positions in K-12 education. Phase three will implement demonstration programs to place Ph.D.s in classrooms and possibly other educational positions in several test states. Phase three will also include an evaluation component to determine the effectiveness of the recruitment effort, as well as the potential benefits to the children taught by Ph.D. teachers.

The charge to the phase one committee was very narrowly directed. We were asked to determine (1) the likelihood that science and mathematics Ph.D.s could be attracted to secondary education, and (2) what special incentives might be useful for states, school districts, and others to attract Ph.D.s to such positions. The committee was not asked to examine or to substantiate the premises that underlie this project, nor was it charged with implementing its findings. Moreover, it was not charged with assessing the potential benefits of placing science and mathematics Ph.D.s in secondary school teaching. These suggested benefits and their cost should be made explicit and carefully evaluated by the phase-two study committee. The charge for the phase one committee was primarily to provide information to the committee overseeing the second phase of the project as it deliberates how demonstration programs might be designed.

METHODS OF STUDY

To meet its charge, the committee investigated the career ambitions of Ph.D.s in the physical sciences, life sciences, and mathematics, and their willingness to take positions in secondary science and mathematics education under a variety of hypothetical conditions. Through focus groups and a national survey of graduate students and postdoctoral fellows, the committee investigated how teacher preparation programs, work conditions, and compensation packages could be modified to attract Ph.D.s to secondary school science and mathematics education.

In addition, the committee interviewed high school and magnet school principals, school district superintendents, state education policy makers, and graduate school deans to identify obstacles in the way of Ph.D.s taking secondary school positions as well as programmatic changes that could be used to attract Ph.D.s to secondary science and mathematics education. The committee also conducted interviews with Ph.D.s already working in secondary education to understand any barriers they had to overcome in taking these positions and their experiences in the secondary school environment. Committee members included experts in the life sciences and physical sciences, labor economics, graduate education, secondary school science teaching, teacher preparation, and partnerships between higher education and K-12 education institutions.

FINDINGS

Interest in Secondary Teaching

Based on our survey results, a large enough number of Ph.D.s appear to be sufficiently interested in secondary education for the NRC to explore a program to attract Ph.D.s to secondary school teaching positions. However, we cannot estimate the exact percentage of Ph.D.s who might ultimately consider secondary school teaching as a career. As with any career choice, this would depend on the specific incentives offered and the alternatives available at the time of choice. Our survey results demonstrate that potential interest in careers in secondary school science and mathematics education is much higher than the 0.8 percent of Ph.D.s who currently work in K-12 education. The interest is high enough, we believe, to justify the development of demonstration programs to test the feasibility of this career alternative. This group of highly trained and knowledgeable individuals is potentially a valuable resource for secondary school science and mathematics education.

Respondents to our survey have typically considered at least four different options in contemplating their career futures; however, at least 36 percent of respondents have considered secondary school teaching or other secondary education positions in their career decision-making. Respondents who were still in graduate school, female, or U.S. citizens were the most open to considering a career in secondary education. Chemists, with strong career options in industry, were less likely than respondents in the biological sciences, physics, and mathematics to consider secondary teaching positions.

Challenges and Possibilities

Given our survey results, a key question that a project designed to attract Ph.D.s to secondary school teaching must address is why, if up to 36 percent of science and mathematics Ph.D.s have considered secondary teaching careers, less than 1 percent are currently employed by K-12 educational institutions. We found that Ph.D.s have many negative perceptions about secondary school education that mitigate against their considering secondary school teaching positions. They perceive a lack of status and respect as teachers, poor classroom laboratory facilities, too many students in classrooms, structured curricula with little opportunity for creativity, possible conflicts with non-Ph.D. teachers, and student discipline problems. They also often perceive little value in education courses and see teacher certification as a barrier that is difficult to overcome. Low salary expectations for teaching in comparison to other careers also present a significant disincentive.

Stereotypes about Ph.D.s both in the secondary schools and in the universities create obstacles. For example, many school administrators argue that Ph.D.s may have good content knowledge, but do not have necessary pedagogical skills or cannot relate to secondary school students. In addition many university faculty do not promote non-academic careers for Ph.D.s, much less careers in secondary school education. Indeed, graduate students typically aspire to positions in academic science and mathematics similar to those of their mentors, and the socialization process in graduate school strongly reinforces this career path.

Given these challenges, what do we, the phase one committee, believe is necessary for success? First, a program to attract Ph.D.s to secondary school teaching must combat negative perceptions about the secondary school environment. A first step would be to recruit Ph.D.s for whom the perceived positives outweigh the negatives. This may not be so difficult. While many participants in our focus groups held negative perceptions, many also held a number of positive perceptions, which included attractive working hours, a work schedule similar to their children's school day, and time for research or other activities during the summer. Many also believed they would enjoy the opportunity to foster the scientific interests of young people. Differences among focus group participants about whether they would prefer to teach at regular public high schools or science and technology magnet schools suggest that flexibility vis-à-vis these preferences will be important if we want to attract Ph.D.s to teaching.

Our interviews with Ph.D. teachers and school administrators indicated that negative stereotypes about Ph.D.s as teachers are widespread, but have not posed obstacles to all Ph.D.s who have actually become teachers. In practice, only a minority of the Ph.D. teachers we spoke with had encountered resistance from school administrators or teachers. We also found that only a minority of the teachers we interviewed faced active disregard from colleagues and mentors after announcing a decision to take a secondary school position. We suggest that focusing on those states, school districts, schools, and graduate institutions where individuals—faculty and administrators—are most supportive of secondary school education as a potential career path for Ph.D.s would provide the most fertile ground for the demonstration projects.

We learned that those who select a secondary school teaching career should first assess their own personalities, interests, and skills. Those who will succeed and find fulfillment in secondary school education will be those who love teaching and enjoy helping students learn and achieve. Ph.D. teachers told us that for them the love of teaching and the enjoyment they get from working with children helps compensate for the higher salaries they could command in other science and mathematics careers. They also told us that preparation in educational pedagogy is essential, even for Ph.D.s, and that certification is an important outward sign of professional acceptance. Finally, it appears important that Ph.D.s contemplating teaching should have had some prior relevant experience and education to help them determine whether a teaching career is right for them.

Our data suggest that Ph.D.s can be attracted to secondary school education through programs that address their needs and interests and that help sustain them as teachers. Our survey presented graduate students and recent Ph.D.s with a number of scenarios, under which they were asked if they would consider secondary school science or mathematics teaching. Respondents indicated that they would be attracted by a fellowship program that provided training, placement, opportunities for networking with peers and that would cover their living expenses during the training period. They were strongly disinclined to undergo a normal full certification for teaching, but were quite amenable to an accelerated program. They were also interested in receiving mentoring during their classroom training. The potential availability of better resources for science education in the classroom and of better salaries were also of considerable interest to our survey respondents.

Other incentives that would serve to attract Ph.D.s to teaching careers would be access to a regional- or university-based science teaching resource center that provided science kits, loaned laboratory equipment, and organized field opportunities for science experiments in which students could participate. Many Ph.D.s would like to continue their involvement in science in some way. Our survey respondents who said they would consider secondary school careers indicated that funding for summer research opportunities and for attendance at professional meetings during the school year is very appealing to them.

Although many survey respondents say they would be unwilling to make more than an initial two-year commitment to secondary school education this should not be seen as limiting any activity designed to attract Ph.D.s to such positions. We believe it important first to recruit Ph.D.s to secondary education and then to work to retain them, as any industry tries to retain its valued employees. We anticipate that some of these individuals will discover their love for teaching children and remain in the program far beyond the initial two-year commitment. Among incentives for retention might be special training opportunities that would facilitate career options in K-12 leadership roles.

GUIDANCE FOR DEMONSTRATION PROGRAMS

As a next step, we recommend that the NRC continue to explore the development of demonstration programs. We recommend that the committee overseeing the second phase of the NRC's project on attracting Ph.D.s to secondary science and mathematics education convene a workshop consisting of stakeholders in secondary school and postsecondary science and mathematics education to discuss in detail how demonstration programs might be structured. Based on the findings of the phase one study, we urge the committee overseeing phase two to consider the following programmatic features for demonstrations programs.

ORGANIZATION OF DEMONSTRATION PROGRAMS

State Demonstration Programs. The committee overseeing the second phase of this project should consider developing demonstration programs in cooperation with a small number of interested states. State governments should organize these demonstration programs because states play a stronger role than the federal government in education in the United States and can potentially bring more resources to bear than can local school districts. States may also develop their demonstration programs to fit their own educational and human resources needs.

Selection and Placement of Ph.D.s for Teaching Positions. The committee recognized that particular care must be taken in selecting Ph.D.s to participate in these demonstration programs. The selection process should identify individuals who have strong knowledge of their subject matter, a demonstrated interest in secondary school science and mathematics education, and personal characteristics appropriate to the secondary school education environment.

Drawing from survey results, the committee suggests that state demonstration programs place and support Ph.D.s in a variety of secondary school education positions, including teaching positions in regular public secondary schools and science and technology magnet schools, as appropriate to the needs of the state. States should consider regional clustering of Ph.D.s in their demonstration programs to facilitate networking, to optimize use of laboratory and science teaching resources, and to forge links between demonstration programs and university education science departments.

Role of Postsecondary Institutions. The phase two committee should consider designing demonstration programs that have strong linkages to science and mathematics programs at colleges and universities in their states, piggybacking on any existing partnership programs. Colleges and universities could facilitate the recruitment and preparation of Ph.D.s for secondary school teaching and provide opportunities for classroom and secondary school experience for Ph.D.s interested in applying to the demonstration programs. They could also serve as venues for special workshops and meetings for Ph.D. teachers during the school year as part of a demonstration program in a given state. Finally, they could provide resources to support secondary school science and mathematics education.

Evaluation. Finally, we suggest that any demonstration program designed by the phase two committee include an evaluation component, to be implemented simultaneously with the demonstrations. The evaluation plan should address the feasibility of placing science and mathematics Ph.D.s in secondary school teaching, assessing the process of implementing such a program, and conducting an outcome evaluation based on measurable goals. A cross-site evaluation of the state demonstration programs, including their means for recruiting, placing, and supporting Ph.D.s in secondary school teaching, would inform other states considering similar programs.

PEDAGOGICAL SKILLS AND TEACHING RESOURCES

Education Courses and Certification. We strongly support the development of education courses and a teacher certification process tailored to the experiences and needs of Ph.D. scientists and mathematicians. Interviews with administrators and Ph.D. teachers indicated that education courses provide teachers with important pedagogical knowledge and that certification is an important step in establishing oneself as a teacher. We strongly agree. As a practical matter, however, courses leading to certification are not likely to be attractive to this population unless they can be accomplished in a fairly compressed manner. We found that 44 percent of our survey respondents and more than two-thirds of those who had previously considered teaching careers indicated that they would consider teaching positions if they could receive their main training prior to beginning teaching by taking an intensive summer course in education. The percentage who would consider teaching if the period of time were increased to one year dropped precipitously to just 14 percent overall and 22 percent for those who had previously considered teaching. Ph.D. teachers we interviewed indicated that they believed a streamlined course in educational theory and practice leading to certification could be developed for Ph.D.s.

We suggest that the state demonstration programs being designed by the phase two committee provide Ph.D.s with an intensive summer program in educational theory and practice as part of a process by which Ph.D.s could obtain teaching certification in an accelerated manner and should fund participants during this summer program. The summer program should focus on educational psychology, pedagogy, and pedagogical content knowledge.

Mentoring and Other Resources. Our survey clearly indicated that Ph.D.s would be more likely to consider teaching positions if they were mentored. The committee suggests that states should consider the selection and appointment of master teachers to serve as mentors to Ph.D.s participating in the demonstration programs. Providing mentors may add programmatic costs if states provide additional compensation to mentors, but the availability of mentors could be an important part of any program to introduce Ph.D.s to teaching careers.

We also found that 52 percent of survey respondents and three-quarters of respondents who had previously considered teaching, would consider a secondary school teaching career if they received support from a regional- or university-based science resource center that provided science kits and loaned equipment or a partnership with an university. The committee overseeing the second phase of this project should work with states to determine whether the development of such science teaching resource centers would be a feasible component of state demonstration programs.

Future Positions for Ph.D.s. Ph.D.s could eventually contribute not only as teachers in the classroom, but also as leaders in other K-12 science and mathematics education positions. There was considerable interest among our survey respondents in providing professional development (e.g. teaching science or mathematics teachers), in becoming a science or mathematics specialist for a school district, in working in a university- or industry-based science educational partnership, or in serving as a science specialist in a science resource center. To a lesser degree, there was also interest in curriculum development or work in a science museum, environmental science center, or similar institution. While Ph.D.s could eventually contribute as leaders in K-12 science and mathematics education through these positions, we believe that it is essential for them to have secondary school teaching experience first.

Incentives

The survey identified a number of incentives that respondents indicated would favorably effect their consideration of taking a position in secondary school teaching.

National Fellowship Program. Two-thirds of our survey respondents and almost 90 percent of respondents who had previously considered secondary school careers would consider taking a position as a secondary school teacher if they were awarded a fellowship that provided training, placement, and special opportunities for networking with peers, and covered living expenses during the training period. Given the potential such a fellowship might have for attracting Ph.D.s to secondary school teaching, the phase two committee should consider ways in which a fellowship program might be established and administered by a prestigious national agency or organization. The national program, instituted in cooperation with the states, could select and train fellows, fund them during their training, and provide an on-going opportunity for networking with peers.

A program that is national in scope could potentially attract resources from national sources, draw applicants from across the country, and serve as a catalyst for the state demonstration programs. A prestigious fellowship program would attract applicants who might not otherwise consider secondary school positions and produce a cohort of science and mathematics teachers who could conceivably change the way science and mathematics are taught. The potential downside to a "prestigious" fellowship for Ph.D. teachers is that it might adversely differentiate them from the population of teachers we want them to join. We also recognize that the establishment of a national fellowship program would increase costs.

Compensation. Survey respondents recognized that salaries for secondary school teaching were lower than for other career options. Still, the average starting salary for teachers anticipated by graduate students and Ph.D.s in our survey—$37,400—is within the range of starting salaries offered to Ph.D. teachers by school districts, albeit at the high end of the range. We believe that states and school districts will need to demonstrate a strong financial commitment to the program for it to succeed by supplementing Ph.D. salaries. This might be done by providing stipends for attendance at scientific meetings and for other activities related to the professional development of the Ph.D.s and the benefit of their students.

We asked our survey respondents if they would consider a secondary school teaching position if they were guaranteed a two-year postdoctoral research fellowship at the end of a two-year teaching position, or if one year of their student loans were forgiven for each year of employment in a full-time teaching position. Given the low favorable response to these scenarios and the additional cost burden that they would place on a national program, we do not recommend that such incentives be offered as part of a national fellows program.

Peer Networking. Survey respondents indicated that, as teachers, they would welcome the opportunity to continue to network with their professional peers. Those surveyed responded very positively to consideration of teaching if they were provided opportunities for networking with peers. A fellowship program could include, among other devices, an annual meeting of fellows; participating states should also consider a regional clustering of Ph.D.s to facilitate networking opportunities.

Connections with the Larger Scientific Community. In designing state demonstration programs the phase two committee should consider providing opportunities for interactions between Ph.D. teachers and the scientific community in academia and industry. Respondents were very likely to consider teaching if they were given funding and time to attend at least one scientific meeting during the school year. We also found that respondents would consider teaching if they were guaranteed a summer fellowship, with travel expenses, in a research laboratory. The phase two committee should consider how states could develop links to universities and businesses to provide summer research opportunities for Ph.D.s, as they already do for science teachers in some states. Finally, the scientific community will need to provide these Ph.D.s with support and treat them as colleagues throughout their careers.

Appendix B
Agenda and Participants

Workshop on Attracting PhDs in Science and Mathematics to Careers in K-12 Education

AGENDA
Aljoya Conference Center
Seattle, Washington
June 2-4, 2000

FRIDAY JUNE 2, 2000

5:15 pm	Welcome	M. Patricia Morse
5:20 pm	Introduction	Bruce Alberts
5:30 pm	Introduction of Committee Members and Logistics	M. Patricia Morse Kevin Aylesworth
7:30 pm	Overview of Committee Charge	M. Patricia Morse
8:00 pm	Data from the Phase 1 Committee on Attracting Science and Mathematics Ph.D.s to Secondary School Teaching	N. Ronald Morris

SATURDAY JUNE 3, 2000

Morning Session

8:30 am   Introduction to Agenda, logistics   M. Patricia Morse
                                                  Kevin Aylesworth

8:45 am   TOPIC 1. Postdoctoral K-12 Career
          Pathways – One major speaker and
          two panel respondents. (Moderator: Kristina Peterson)
             Nancy Hutchison
             Stephanie Shipp
             David Vannier

10:00 am TOPIC 2. Learning, Teaching, Pedagogy and the
          Discipline – Postdoctoral Experiences for the K-12
          Education Environment – One major speaker and
          two panel respondents. (Moderator: Vicki Jacobs)
             Ellen Doris
             Gerhard Salinger
             Mary Long

11:15 am Breakout Groups on Topic 1 and Topic 2

          BREAKOUT TOPIC 1.
          Career Pathways in the Schools
          (Moderator: N. Ronald Morris)
          Career Pathways Informal Venues
          (Moderator: Myles Gordon)

          BREAKOUT TOPIC 2
          Ingredients for "Content" (Moderator: Kimberley Tanner)
          Ingredients for "Content" (Moderator: Margaret Cozzens)

APPENDIX B   65

Afternoon Session

1:30 pm   TOPIC 3. State Opportunities and the Infrastructure –
            One major speaker and two panel respondents.
            (Moderator, David Kennedy)
                Michael McKibbin
                Calvin Frazier
                Francis Eberle

3:00 pm   Breakout Groups – TOPIC 3.

            Needs of States - Collaborations
            (Moderator: David Kennedy)
            Certification and Employment Issues
            (Moderator: Maureen Schifflet)
            Practicum and Mentoring Issues in States
            (Moderator: Arthur Eisenkraft)
            Internships and Summers, Year 2 – State Concerns
            (Moderator: Danine Ezell)

4:00 pm   Synthesis Plenary Session

6:00 pm   Bus to Pacific Science Center

            Reception, Pacific Science Center

SUNDAY JUNE 4, 2000

8:30 am   Summary of Day One Outcomes   M. Patricia Morse

9:00 am   Breakout Sessions:

            1. Filling in the Visions—Issues and Concerns—
               Where Are We Now?

- Cohort I. State Structures – frameworks (Moderator: Myles Gordon)
  - What partnerships with colleges or universities?
  - What formal and informal opportunities for paid positions?
  - Needs of states.
  - What needs to be in place?

- Cohort II. District/School - teacher mentoring – within states (Moderator: Kristina Peterson)
  - What will be the structure in schools?
  - What are appropriate state/university interactions with their cohort of say 5-10 students?
  - What are the rewards for the district master teacher mentor?

- Cohort III. Format of Certification Year (Moderator: Emily Feistritzer)
  - Bring cohort together nationally? How might that be done?
  - State/university partnerships for postdoctoral K-12 mathematics and science pedagogy
  - Issues of collaboration between science disciplines and education colleges
  - Are there many ways to certification? Are some more suitable for postdoctoral students? How can that be addressed?

- Cohort IV. Disciplines and Discipline-based Pedagogies – what is needed and when? (Moderators: Vicki Jacobs and Margaret Cozzens)
  - Mathematics (Moderator: Eric Robinson)
  - Biology/life sciences (Moderator: Angelo Collins)
  - Physical sciences and earth sciences (Moderator: James Stith)

10:00 am   Coffee Break

10:30 am   Breakout Sessions (continued)

       2. Filling in the Visions – Guidance for constructing template(s) around topic

11:30 am   Plenary Reports from three cohorts above
       15 minute report from each cohort and general discussion

APPENDIX B 67

2:00 pm   Breakout Sessions

   3. Filling in the Visions – Issues & templates

- Cohort V. The National Cohort – details and outcomes Recruitment Procedures (Moderator: Danine Ezell)
  - Areas
  - Diversity
  - Years from doctorate
  - Balance in cohort
  - Meeting as a whole – a national college of teacher scholars

- Cohort VI. Meeting the Need to Stay Connected to Research (Moderator: Maynard Olson)
  - How does the postdoc maintain ties with the discipline?
  - Research in teaching the content?
  - Professional societies – delivering scholarly education papers
  - Summer internships – in discipline research

- Cohort VII. Science Discipline Connections and Mentoring (Moderator: James Stith)
  - National discipline mentors
  - Role of discipline societies
  - Role of education societies
  - Role and expectations of postdoctoral students at national level

- Cohort VIII. Financial and other Support Considerations (Moderator: MargaretCozzens)

  - What are the needs?
  - How might local, state and national groups participate?
  - How might this be approached?

3:15 pm   Plenary Session Reports of Cohorts
          Final Thoughts and Future Directions   M. Patricia Morse

5:00 pm   Adjourn

PARTICIPANTS

Sigmund Abeles, Project Director, Connecticut Academy Science Assessment Project, Connecticut Academy for Education in Mathematics, Science and Technology, Inc., CT

Howard Adams, H. G. Adams & Associates, Inc. Marietta, GA

Kevin Aylesworth, Senior Program Officer, National Research Council, Center for Education, Washington, DC

Marilyn Baker, Associate Executive Director of the Office of Scientific and Engineering Personnel (OSEP), National Research Council, Washington, DC

Joan Baratz-Snowden, Director of Education Issues Department, American Federation of Teachers, Washington, DC

James Bishop, Associate Professor, Ohio State University, School of Teaching and Learning, Columbus, OH

Dana Riley Black, Associate Director, University of Washington, K-12 Institute for Science, Math, and Technology Education, Seattle, WA

Elizabeth Chatman, Academic Coordinator, University of California, San Francisco, Science & Health Education Partnership, San Francisco, CA

Angelo Collins, Professor of Science Education, Vanderbilt University, TN

Margaret Cozzens, Vice Chancellor for Academic and Student Affairs, University of Colorado, Denver

Tom DeVries, Science Teacher, Vashon Island High School, Burton, WA

Diane Doe, Teacher, San Franciso, CA

Ellen Doris, Author, *Doing What Scientists Do: Children Learn to Investigate Their World*, Colrain, MA

Helen Doyle, Program Officer, The David and Lucile Packard Foundation, Los Altos, CA

Francis Eberle, Executive Director, Maine Mathematics and Science Alliance, Augusta, ME

Karin P. Egan, Program Officer, Carnegie Corporation of New York, Education Division, New York, NY

Arthur Eisenkraft, President, National Science Teaching Association, Arlington, VA; Bedford Public Schools, Bedford, NY

Danine Long Ezell, Science Teacher, The Preuss School UCSD, La Jolla, CA

Emily Feistritzer, President, Center for Education Information, Washington, DC
Calvin M. Frazier, Commissioner of Education, State of Colorado (retired)
Maria Lopez Freeman, Executive Co-Director, State of California, California Science Project, Monterey Park, CA
Bruce Fuchs, Director, National Institutes of Health, Office of Science Education, Rockville, MD
Myles Gordon, Director of Education, American Museum of Natural History, New York, NY
Leslie Guterman, Elementary School Science Specialist, Fremont, CA
Terry Holmer, Senior Project Assistant, National Research Council, Washington, DC
Leroy E. Hood, Gates Professor and Chair, University of Washington, Molecular Biotechnology, Seattle, WA
Richard Hudson, Executive Producer for Science, KTCA, St. Paul, MN
Nancy Hutchison, Hutch Lab, Fred Hutchinson Cancer Research Center, Seattle, WA
Vicki Jacobs, Associate Director, Harvard Graduate School of Education, Harvard Teacher Education Programs, Cambridge, MA
David Kennedy, Director, Instructional Design, Washington State Board of Education, Olympia, WA
Carole Kubota, University of Washington at Bothell, Department of Education, Bothell, WA
Valerie Logan, Outreach Education, Molecular Biotechnology, University of Washington, Seattle, WA
Mary Long, Coordinator of Uteach, University of Texas, Austin, Special Projects Office, Austin, TX
Victoria May, Outreach Director, Washington University, Department of Biology, St. Louis, MO
Michael McKibbin, Consultant, Commission on Teacher Credentialing, Program Evaluation & Research, Sacramento, CA
N. Ronald Morris, Distinguished Professor of Pharmacology, Robert Wood Johnson Medical School, Department of Pharmacology, Piscataway, NJ
Carolyn Morse, Laboratories Manager/Teaching Labs, University of North Carolina, Chapel Hill, Department of Chemistry, Chapel Hill, NC

M. Patricia Morse, Professor (Acting), Department of Zoology University of Washington, Seattle, WA

Richard O'Grady, Executive Director, American Institute of Biological Sciences, Washington, DC

Maynard V. Olson, Professor, Genome Center, University of Washington, Seattle, WA

Karen Peterson, Continuing Education Coordinator, University of Washington, Astronomy Department, Seattle, WA

Kristina Peterson, Chemistry and Biology Teacher, Lakeside School, Seattle, WA

George Reinhart, Program Officer, Office of Scientific and Engineering Personnel (OSEP), National Research Council, Washington, DC

Eric Robinson, Associate Professor and Director of COMPASS, Ithaca College, Department of Mathematics and Computer Science, Ithaca, NY

Karolyn Rohr, Coordinator, Resident Teacher Program, Montgomery County Public Schools, Department of Human Resources, Rockville, MD

Joan Rothenberg, Office of Congressman Rush Holt, Washington, DC

Gerhard Salinger, Program Director, Division of Elementary, Secondary, and Informal Education, National Science Foundation, Arlington, VA

Dennis Schatz, Associate Director of Education, Pacific Science Center, Seattle, WA

Maureen Shiflett, Education Consultant, Buena Park, CA

Ray Shiflett, Professor of Mathematics, California Polytechnic University, Pomona, CA

Stephanie Shipp, Research Associate/Faculty Lecturer, Rice University, Department of Geology and Geophysics MS126, Houston, TX

Mary Grace Snyder, Coordinator, Resident Teacher Program, Montgomery County Public Schools, Department of Human Resources, Rockville, MD

James H. Stith, Director of Physics Programs, American Institute of Physics, College Park, MD

Gerald M. Stokes, Associate Lab Director, Environmental and Health Science Division, Battelle, Pacific Northwest National Lab, Richland, WA

Kimberley Tanner, Post doc in Science Education, University of California, San Francisco, Science and Health Partnership, San Francisco, CA

Jan Tuomi, Science Consultant, Eisenhower Regional Consortium, Denver, CO

David Vannier, American Associaton for the Advancement of Science/National Science Foundation Science and Engineering Fellow, Arlington, VA

Appendix C

Nontraditional K-12 Teacher Preparation Programs

This appendix describes three programs that offer nontraditional ways for both college students and midcareer professionals to become K-12 teachers: UTeach, at the University of Texas at Austin, the midcareer math and science program at the Harvard Graduate School of Education, and Teach for America. This appendix discusses their history, candidate selection process, and some other features. As noted in Chapter 3, none of these is a program to train new and recent PhDs for K-12 teaching, but they do offer ideas about approaches to consider in designing a program for potential teachers.

UTEACH

History and Goals

UTeach is an innovative program at the University of Texas at Austin that prepares science and mathematics majors to teach in secondary schools (see http://www.uteach.utexas.edu). The program, which began in fall 1997 with 28 students, involves a partnership of the College of Natural Sciences, the College of Education, and the local school district.

The program was developed after the dean of the College of Natural Sciences, in response to the national shortage of highly qualified science and mathematics teachers, surveyed students in the college. The college has an enrollment of more than 8,000 students majoring in science, mathematics, or computer science. The results revealed that more than one-third of the students had considered teaching as a career, although only small numbers of students were pursuing teaching certification from the university during the 1990s. Concerned that this potential population of future teachers was not being tapped, the dean called on three outstanding secondary school teachers from the local school district and an assessment expert to propose an ideal teacher training program in science and mathematics. During the summer of 1997, this group developed a framework for a 4-year course of study, taking into account university regulations, and paying close attention to recently modified state and national guidelines. This plan was reviewed by interested professors and deans in both the College of Natural Sciences and the College of Education and also sent to education leaders across the state and representatives of the State Board for Educator Certification for their review and input.

The document that resulted was a blueprint for UTeach. It featured early guided field experience, graduation in 4 years with a degree from the College of Natural Sciences and a recommendation for state certification from the College of Education. The program is based on a three-fold partnership, drawing on the experience of master teachers from the school district, the instructional and curricular knowledge of faculty in the College of Education, and the content-area strength of scientists and mathematicians in the College of Natural Sciences. The UTeach Program is designed to be flexible and to accommodate diverse student schedules.

Candidate Selection and Enrollment

A distinctive feature of the UTeach Program is that it does not have a formal selection process. The nature of the program's first year is to encourage all mathematics and science majors at the University of Texas at Austin to explore whether the teaching profession is a career option for them. The College of Natural Sciences invites freshmen, sophomores, juniors, and seniors to join the program, and it offers free tuition for the first two, 1-hour courses. To the extent that the university capacity allows, it admits a select group of post baccalaureates into the program.

In spring 2001 there were 280 students in the program, and the first major group of 35 students graduated in May 2001. UTeach is one of the largest programs of this type at any research university in the country: Its enrollment by fall 2001 was expected to be more than 300, and it is projected that the following year the program will reach a steady state of about

500 students a year. This means that the university will produce about 100 new mathematics and science teachers each year.

Classroom Teaching Experiences

In accordance with new national guidelines for teacher preparation that recommend early field-based experience, UTeach students enrolled in the introductory Step classes begin carefully supervised classroom teaching in local elementary school classrooms during their first semester in the program. Entering students begin with Step 1, the introductory field-based course that places preservice teachers in elementary classrooms. They then move on to Step 2, the second field-based course that offers UTeach students a chance to teach at the middle grade level. Working with excellent mentor teachers trained to supervise them with oral and written evaluations, UTeach students are encouraged to discover as early as their freshman year whether or not they are truly interested in teaching as a career. The classroom practice is usually an exciting and positive experience that raises the level of a student's commitment to teaching. This early field-based experience is sustained throughout the course sequences, enabling students to gain exposure to elementary, middle, and high school students and to the diverse local population.

Field-based experiences take place primarily in inner-city schools with high minority student populations with low socioeconomic status. They expose the UTeach students to the challenges of teaching in an urban setting and to an awareness of the difference that enthusiastic teachers can make to their students. This arrangement expands the schools' capacity to meet the needs of disadvantaged children, and it encourages UTeach students to pursue teaching jobs in high-need, inner-city settings.

Coursework

The College of Natural Sciences has developed 16 new degree plans for UTeach that lead to undergraduate degrees in such specializations as mathematics, computer sciences, biological sciences, chemistry, physics, as well as Texas state certification for teaching secondary students. Faculty in the College of Education have developed a completely new set of professional development courses for UTeach. In order to foster interdisciplinary teaching in our students, students participate in professional development courses together. In contrast with a traditional menu of methods and observation courses, the new professional development sequence is specifically focused on the challenges of learning secondary science and mathematics. The courses give students a foundation in the theory of learning, but they also challenge the students' knowledge of science and mathematics content areas. The courses combine educational theory with practical applications, current instructional technologies, and field experiences.

Certification

Under current state guidelines, UTeach students are certified to teach in grades 6–12. Most of the students obtain composite certification, which enables them to teach all of the science disciplines. Starting in 2003, the program will respond to new state categories for certification: students will obtain certification for grades 4–8 (middle school) or 8–12. UTeach will be expanded and modified to provide the middle grades certification option.

THE MIDCAREER MATH AND SCIENCE PROGRAM

History and Goals

Established in 1983, the Harvard Graduate School of Education's MidCareer Math and Science (MCMS) Program was designed to address the shortage of well-qualified, secondary teachers in mathematics and the sciences specifically by providing a venue for mid-career professionals from non-educational but technically oriented fields to enter careers as secondary science and mathematics teachers. (See http://www.gse.harvard.edu/~admit/progs-edm-tac-overview.html for more information.) The program is based on the premise that these new teachers can enrich secondary school environments and curricula with knowledge and materials drawn from their professional and practical experiences. Over the years, the MCMS Program has served as a national model for other universities and for state legislatures to consider for adoption.

The objectives of the MidCareer Math and Science Program include:

- to attend to the condition of pre-collegiate math and science education by crafting an innovative response to the need for qualified new teachers and to the need for professional development of experienced professionals;

- to attract a generally underutilized labor pool—mid-career professionals—for the education our nation's youth;
- to rekindle and revitalize an intense commitment, interest, and pride in the work of practicing educators; and
- to provide an innovative model for other educational institutions across the country to address the condition of secondary math and science education.

In addition, the MidCareer Math and Science Program is designed to address the larger goals of the Harvard Teacher Education Program, including:

- to prepare teachers for the specific challenges of urban education including providing high quality instruction for all students, addressing the causes of unequal access in our educational system, and creating classrooms and schools where previously unsuccessful students can succeed;
- to prepare teachers to facilitate students' understandings and capacities to construct knowledge through a deep understanding of literacy and to assume new leadership roles within schools by developing skills to participate in organizational diagnosis and change leadership activities; and
- to demonstrate that teaching can be a life-long career with multiple stages and aspects of growth typical of adult development by starting novice teachers on a path toward National Board Certification.

In carrying out these objectives, the MCMS Program works closely with cooperating urban middle and high schools in preparing its candidates for teaching. Extensive fieldwork in secondary school classrooms is the centerpiece of the graduate studies of MCMS candidates. The MCMS Program shares with HGSE a concern about the integration of practice and theory, as well as of curricular design and pedagogy, combining field-based experience with critical reflection on the nature and purposes of teaching and learning. In addition, MCMS supports it candidates in making use of their previous professional experience to develop real-world applications in secondary-school curriculum.

Candidates who successfully complete the MCMS Program receive a masters in education and are eligible for middle- and/or secondary-school certification in the Commonwealth of Massachusetts in the appropriate content (i.e., biology, chemistry, earth science, general science, mathematics, or physics). Since the inception of the MCMS Program during the

1983-1984 academic year, it has graduated about 300 candidates. The diverse program participants have included a retired Rear Admiral; electrical, computer, and civil engineers; biochemistry and microbiology specialists; physicists; statisticians; a meteorologist from the U.S. Weather Service; NASA affiliates; bankers; members of the U.S. Diplomatic Corps; a veterinary technician; and geologists.

Candidate Selection

To have been eligible for admissions to the program for 2001-2002, applicants had to have completed a major in their chosen teaching field or the equivalent of a major (having no more than two courses to fulfill to meet Massachusetts' subject-matter requirements). They were required to submit an application, all transcripts of post-secondary study, three references, a personal statement (in response to questions specifically related to teaching and learning), and scores on either the Graduate Record Exam or the Miller Analogy Test (all of which are required for general admission to the Harvard Graduate School of Education). The Admissions Committee was composed of faculty, current Harvard students, and practitioners.

Classroom Teaching Experiences

Extensive fieldwork in secondary school classrooms is the centerpiece of the graduate studies of MCMS candidates. MCMS candidates complete their field-based requirements in a public high or middle school, working with experienced, state-certified teachers in urban schools within the greater Boston area that belong to Harvard's Professional School Partnership.

For the 2001-2002 academic year, MCMS candidates completed a minimum of a 75 clock-hour pre-practicum (observing, assisting, co-teaching, and/or teaching) and a minimum of a 150 clock-hour practicum (undertaking clearly defined, supervised instructional responsibility) to fulfill requirements of provisional certification with advanced standing. During this time, the program also encouraged candidates to observe in other classrooms (both within and without their own departments) and in at least one school that was substantially different from their own placement site. Interns completed their pre-practica and practica requirements in the fall term, enrolling, at the same time, in course work designed to address their fieldwork experiences and to bridge theory and practice.

During the 2001-2002 academic year, MCMS candidates completed a spring-term, semester-long clinical experience (of at least 400 clock-hours) to fulfill the field-based requirements for standard certification. During the clinical experience, candidates assumed teaching responsibility for the equivalent of three 45-50 minute classes in their chosen area of certification. In addition, the program asked that candidates work with guidance counselors, extra-curricular activities supervisors, and special-needs teachers and consult with their Teacher Mentors on a regular basis. During their clinical experiences, candidates enrolled in a seminar designed to support their systematic reflection on their implementation of curriculum, instruction, and assessment.

Coursework

MCMS candidates complete the equivalent of nine courses to fulfill requirements for a masters in education and state certification.

Summer 2001
- "Adolescent Development and Identity Formation" (module)
- "Introduction to Teaching" (module)
- "Summer Practicum" (module)

Fall 2001
- "Equity and Education"
- "Teaching English, History/Social Studies, or Mathematics/Science"
- "Literacy and Learning" (module)
- One elective

Spring 2002
- "Teacher Leadership and Research" (module)
- "School Reform: Curriculum and Instructional Leadership"
- "Urban Education Seminar" (module)
- One elective

Year-long
- "Practicum/Clinical Experience in Secondary Education"

Outcomes

Generally, all HGSE MCMS graduates who seek teaching positions procure them. At any one time, over the years, about 80 percent of all program graduates are teaching.

TEACH FOR AMERICA

History and Goals

Teach for America began in 1989 as a senior thesis project by Wendy Kopp, who was then a student at Princeton University. The program places college graduates from all academic fields in 2-year teaching positions at public schools in low-income areas across the nation. A goal of the program is to help these districts with few resources to provide a high-quality education to their students. The school districts hire and pay corps members as regular beginning teachers. Local offices help orient corps members to their new schools and communities and foster professional and personal support networks. See http://www.teachforamerica.org/tfa for more information.

The first year, 2,500 men and women from more than 100 colleges applied to become members of what became known as a corps. From the pool of applicants, 500 corps members were chosen. Over the past 12 years, Teach for America has placed 8,000 corps members in 16 urban and rural areas. During the 2000-2001 academic year, corps members were placed at schools in Atlanta, Baltimore, the Bay Area in California, Chicago, Houston, Los Angeles, the Mississippi Delta, New Jersey, New Orleans, New York City, North Carolina, Phoenix, the Rio Grande Valley, rural Louisiana, and Washington, D.C. About 59 percent of the placements were at the elementary level, and 41 percent were at the secondary level.

Candidate Selection

To apply for the program, candidates fill out a written application that includes a letter of intent, resume, and essay. When they are interviewed, applicants indicate their preferences for regional sites, grade levels, and subject areas. The program strives to meet the applicants' preferences while meeting districts' needs. Consideration is also is given to district and state requirements—for example, requirements for the corps members to have a certain number of coursework hours in a subject before they are allowed to teach it.

A selection committee reviews the applications and invites the most promising candidates to participate in a day-long interview, which includes a sample teaching lesson, a group discussion, and a personal interview. The selection committee looks for evidence that applicants have an ability to thrive on overcoming challenges, a drive to achieve results, a commitment to setting only the highest expectations for themselves and their students.

Teach for America attempts to place corps members in schools that employ other corps members or alumni so that they can collaborate on projects and support one another's professional growth. It also provides support for current and past corps members through monthly newsletters, electronic discussion groups, retreats, and social activities. In addition, local program staff works with area school districts, schools of education, professional associations, and other organizations to give corps members access to the best professional development and teaching resources available. At the national level, inter-regional conferences help corps members stay connected to their colleagues across the country.

Teaching Preparation

New corps members attend a preservice summer training institute for 5 weeks in either Houston or New York City, depending on their placement region. During the training program, the corps members learn about and practice approaches used by successful teachers in low-income communities. They also learn basic teaching skills, such as curricular planning around clear goals, effective lesson design, student assessment, classroom management techniques, and literacy development.

In the mornings and early afternoons of the institute, corps members teach in a summer school program. In the afternoons and evenings, they participate in a full schedule of discussions, workshops, and other professional development activities with a faculty of exceptional corps members, alumni, and other experienced educators. The institutes also incorporate ceremonies and special events that are designed to build a sense of community among the corps members and allow them to appreciate the magnitude of their collective effect on the lives of K-12 students.

Near the end or after the institute, corps members travel to their assigned site to participate in a 1- to 2-week local orientation. During this induction period, corps members learn about the local communities in which they will teach, locate housing, and interview for their actual teaching positions.

School districts hire and pay corps members as regular beginning teachers, but use alternate routes to teacher certification. This arrangement allows corps members to teach without the education courses that are typically required of teachers in public schools. In many cases, they must take courses after they begin teaching. Teach for America has formed partnerships with school districts, states, and schools of education to enable corps members to take the coursework they need after they begin teaching.

Outcomes

Of the program's alumni, some of whom were placed as teachers in 1990, 60 percent are still working full-time in education: 40 percent are teaching, and 20 percent are working in other positions in the field of K-12 education. In the latter category, for example, some alumni work in school or district administration, or in organizations that are focused on education reform. Other alumni are graduate students in education.

An independent evaluation of the effect of the program's teachers on student performance in the Houston Independent School District was conducted in 2001 by CREDO, a research group based at the Hoover Institution of Stanford University. The evaluation compared Teach for America participants with all other new teachers recruited during the same years. The evaluation reached three main conclusions (Raymond, M., et al, 2001):

- The effect of having a teacher from Teach for America was generally positive. The positive effect appears to be largest in mathematics, in both elementary and middle school. Results in reading also were positive, but the magnitudes of effects were smaller.
- The differences between the average Teach for America teacher's and other teachers' contribution to students' perforamnce are generally not statistically significant.
- TFA teachers as a group show less variation in quality than teachers entering from different routes.

The evaluation also indicated that the Teach for America teachers were more likely to hold a bachelor's degree, were placed in more difficult classes, and were less likely to leave after the first year than were the other teachers.

Appendix D

Programs to Strengthen Connections Between Professions and K-12 Education

This appendix describes four programs that have been implemented in recent years to strengthen the connections between the science, mathematics, and engineering professions and K-12 education. Three of them are national programs sponsored by the National Science Foundation (NSF); the fourth is a small, local program sponsored jointly by the National Institutes of Health (NIH) and the Montgomery (Maryland) County Public Schools (MCPS). Only the last one is designed to lead to secondary school teaching. However, all three are examples of programs to strengthen the science, mathematics, and technology infrastructure of the nation's schools.

These programs have the potential to play significant roles in the improvement of K-12 education in science, mathematics, engineering, and technology. They are similar to the demonstration program proposed in this report because they target PhDs or graduate students in science, mathematics, engineering, and technology, explicitly support standards-based K-12 science and mathematics instruction, and are predicated on some kind of partnership between higher education and school districts. The MCPS/NIH program had not been evaluated at the time of this report. The goals of other programs are significantly different from the one proposed in this report, and their available evaluation information provided no specific guidance to the committee.

NSF PROGRAMS

The NSF has supported three programs over the last 5 years that are designed to provide pre- and postdoctoral experiences in K-12 education for scientists, mathematicians, and engineers. The programs are the Postdoctoral Fellowships in Science, Mathematics, Engineering, and Technology Education (see NSF, 2001a and http://www.ehr.nsf.gov/dge/programs/pfsmete/), the Graduate Teaching Fellowships in K-12 Education (see NSF, 2001b and http://www.nsf.gov/home/crssprgm/gk12/start.htm), and the Centers for Learning and Teaching (see NSF, 2001c and http://www.ehr.nsf.gov/esie/resources/centers.asp).

Postdoctoral Fellowships in Science, Mathematics, Engineering, and Technology Education

The NSF Postdoctoral Fellowships in Science, Mathematics, Engineering and Technology Education (PFSMETE) have two goals:

- to prepare PhD graduates in science, mathematics, engineering or technology with the necessary skills to assume leadership roles in science education in the nation's diverse academic institutions, and
- to provide opportunities for outstanding PhD graduates to develop expertise in a facet of science education research that would qualify them for the new range of academic positions that will come with the 21st century.

The fellowships, which were first awarded in fiscal 1997, are given to academic institutions. The first awards were used to support 63 fellowships for PhDs in the earth sciences, engineering, chemistry, the life sciences, mathematics, physics, and psychology. The fellows' projects included work in cognition and learning, community-based research, curriculum development, educational technology, evaluation and assessment in K-12 education, and teacher education.

Graduate Teaching Fellowships in K-12 Education

The primary objective of the Graduate Teaching Fellowships in K-12 Education Program (GK-12) is to provide fellowships to highly qualified graduate and advanced undergraduate students in science, mathematics, and engineering to serve as resources in the nation's K-12 schools. GK-12 fellows work directly with teachers on many activities, including:

- demonstrating key concepts in science or mathematics;
- modeling for students the habits and skills needed to pursue future study in science, mathematics, and engineering;
- serving as role models for future science, mathematics, and engineering professionals;
- enhancing teachers' content knowledge and understanding of principles of science and mathematics; and
- assisting in science and mathematics instruction.

Most of the GK-12 projects do not provide the fellows with classroom teaching experience. However, one project—a partnership between Johns Hopkins School of Medicine and Paul Lawrence Dunbar High School—does provide the fellows with formal preparation for classroom teaching and student teaching experience. That project has only four fellows, and it is not explicitly intended as a way for GK-12 fellows to become certified K-12 teachers or to ground them in classroom practice for whatever segment of K-12 education they choose to pursue.

The NSF GK-12 program began as a pilot project in 1999, but NSF received a greater than anticipated number of proposals and increased the funds beyond the level originally planned. The program was subsequently included in NSF's annual budget and approved by Congress. In its first 2 years, the GK-12 program has provided $43.4 million for 56 grants to academic institutions across the country. It was scheduled to allocate an additional $10 million in fiscal 2001. In this popular program, approximately 600 graduate and advanced undergraduate students in science, mathematics, engineering and technology have served as teaching fellows in K-12 schools to date.

Centers for Teaching and Learning

NSF initiated the Centers for Learning and Teaching in fiscal 2001 with three goals (NSF, 2001c):

(1) to increase significantly the numbers of K-12 science, mathematics, engineering, and technology educators in formal (schools) or informal (museums, zoos, botanical gardens, etc.) settings who

have current content knowledge in their disciplinary area and who are prepared to implement standards-based instruction and new assessment strategies;

(2) to rebuild and diversify the human resource base that forms the national infrastructure for science, mathematics, engineering, and technology; and

(3) to provide substantive opportunities for research into the nature of learning, strategies of teaching, policies of educational reform, and outcomes of standards-based reform.

Approximately $16 million was initially allocated for 7-9 centers awards and $2 million for 8-12 awards to develop proposals for future centers.

For the participants—which include PhDs, university teacher educators, curriculum developers, district-level or state-level supervisors and coordinators, lead teachers, informal science educators, assessment specialists, and school administrators (e.g., principals)—the program has a postdoctoral and internship program, which includes graduate programs of study (for MS, PhD, or EdD degrees). Participants in this part of the program are offered a variety of options for developing special expertise. The program description provides the following examples:

One type of Center might focus on developing high quality K-12 science curricular materials and bring together representatives from school districts, informal science centers, curriculum developers, undergraduate, graduate, and post-doctoral students, and science faculty to design, develop, and field-test new materials. Another type of Center might focus on research, evaluation, and assessment through emphasis on the graduate education of educational psychologists and psychometricians who focus on the learning and assessment of mathematics and/or science and who are needed to evaluate large-scale reform projects such as the SME&T [science, mathematics, engineering, and technology] systemic initiatives. Another Center might choose to address the retraining of those who already hold a doctorate (or the equivalent) in science, mathematics, and engineering and who have particular interest in SME&T education (NSF, 2001c, p. 9)

TRAINING TEACHERS FOR TOMORROW

The Training Teachers for Tomorrow program is a partnership between the NIH and the MCPS to help NIH postdoctoral fellows make a transition to careers as certified secondary school teachers. The program began in the fall 2000 with funding under a 6-year grant from the state of Maryland to pay the tuition for the program participants. The PhDs fill teaching vacancies at MCPS in their field of expertise for the school year and receive the salary and benefits of a first-year teacher. MCPS provides mentors for each participant; NIH provides some stipends for opportunities to work in laboratories during the summer. Although the program was designed to meet the needs of NIH postdoctoral fellows, the MCPS part of the program is open to other midcareer professionals. There were 2 postdoctoral fellows in the first cohort of 13, and the program administrators expect an average of 2-4 postdoctoral fellows in the program each year.

Program participants receive orientation training in the summer just prior to the start of the their first classroom teaching experience. They are classified as "resident teachers" until they complete the 2-year course of studies, pass both Parts I and II of the Praxis exam,[1] and receive a successful evaluation for at least 1 year of teaching. Fulfilling these three requirements makes them eligible for a teaching certificate from the state of Maryland. The teacher education courses are offered at an MCPS school site, taught by MCPS master teachers. The integration of coursework with classroom teaching allows the resident teachers to understand the relevance of the coursework to their classroom and to apply what they learn in their teaching practice.

The coursework covers the following areas, as required by the Maryland State Department of Education: human learning, adolescent development, special needs students, assessment, teaching methods, and reading 1 and 2. To address pedagogical skills, the cohort meets weekly with experienced, master MCPS teachers. All preparation activities are tied directly to the participants' teaching activities in their classrooms. Topics are addressed in a way that anticipates their needs, answers their questions, and helps them plan successful instruction.

---

[1] The Praxis exams are standard tests used in many states as one requirement for teacher certification.

Appendix E

Gallup Teacher Perceiver Instrument

The Gallup Teacher Perceiver is a structured interview that consists of a set of open-ended items (see Gallup Organization, 2002 and http://education.gallup.com/select/themeTeach.asp). It is based on Gallup's research on what are believed to be the characteristics that make the best teachers. The interviews may be given face to face or over the telephone. The items covered in the instrument are listed below, with the descriptions provided by Gallup (2002).

MISSION—Mission is what takes some individuals and groups out of society's mainstream in order to assure the quality and purposiveness of that mainstream. Mission is a deep underlying belief that students can grow and attain self-actualization. A teacher with mission has a goal to make a significant contribution to other people.

EMPATHY—Empathy is the apprehension and acceptance of the state of mind of another person. Practically, we say we put ourselves into the other person's place. Empathy is the phenomenon that provides the teacher feedback about the individual student's feelings and thoughts.

RAPPORT DRIVE—The rapport drive is evidenced by the teacher's ability to have an approving and mutually favorable relationship with each student. The teacher likes students and expects them to reciprocate. Rapport is seen by the teacher as a favorable and necessary condition of learning.

INDIVIDUALIZED PERCEPTION—Individualized perception means that the teacher spontaneously thinks about the interests and needs of each student and makes every effort to personalize each student's program.

LISTENING—The Listening theme is evident when a person spontaneously listens to others with responsiveness and acceptance. Listening is viewed as beneficial to the speaker.

INVESTMENT—The Investment theme is indicated by the teacher's capacity to receive a satisfaction from the growth of the students. This is in contrast to the person who must personally perform to achieve satisfaction.

INPUT DRIVE—Input drive is evidenced by the teacher who is continuously searching for ideas, materials and experiences to use in helping other people, especially students.

ACTIVATION—Activation indicates that the teacher is capable of stimulating students to think, to respond, to feel, to learn.

INNOVATION—The innovation theme is indicated when a teacher tries new ideas and techniques. A certain amount of determination is observed in this theme because the idea has to be implemented. At a higher level of innovation is creativity, where the teacher has the capability of putting information and experience together into new configurations.

GESTALT—The Gestalt theme indicates the teacher has a drive toward completeness. The teacher sees in patterns—is uneasy until work is finished. When gestalt is high, the teacher tends toward perfectionism. Even though form and structure are important, the individual student is considered first. The teacher works from individual to structure.

OBJECTIVITY—Objectivity is indicated when a teacher responds to the total situation. This teacher gets facts and understands first as compared to making an impulsive reaction.

FOCUS—Focus is indicated when a person has models and goals. The person's life is moving in a planned direction. The teacher knows what the goals are and selects activities in terms of these goals.

Appendix F

National Residency Matching Program

The following description of the program used to place applicants for graduate medical schools through the National Resident Matching Program (NRMP) is taken from National Resident Matching Program, 2002. Applicants send their ranked list of program to NRMP, and programs send their ranked list of applicants to NRMP.

The NRMP matching algorithm begins with an attempt to place an applicant into the program indicated as most preferred on that applicant's list. If the applicant cannot be matched to his or her first-choice program, an attempt is then made to place the applicant into the second-choice program, and so on, until there is a tentative match or all the applicant's choices have been exhausted. See http://www.nrmp.org for more information.

If the program also ranks an applicant on its rank order list, the applicant can be tentatively matched to a program in one of two ways:

- If the program has an unfilled position, a tentative match is made between the applicant and program.
- If the program does not have an unfilled position, but the applicant is more attractive to the program than another applicant who is already tentatively matched to the program, the applicant who is the less preferred current match in the program is removed from the program to make room for a tentative match with the more preferred applicant.

Matches are tentative because an applicant who is matched to a program at one point in the matching process may later be removed from the program to make room for an applicant more preferred by the program. When an applicant is removed from a previously made tentative match, an attempt is made to rematch the applicant, starting from the top of his or her list. This process is carried out for all applicants until each applicant has either been tentatively matched to the most preferred choice possible or all choices submitted by the applicant have been exhausted. When all applicants have been considered, the process is complete and all tentative matches become final.

Appendix G

Biographical Information on Committee Members

M. Patricia Morse, Chair, is a marine biologist and science educator at the University of Washington, Seattle. For 34 years, she was professor of biology at Northeastern University, and the last 4 she also served as a program director at the U.S. National Science Foundation (NSF) in the Division of Elementary, Secondary and Informal Education, where she served as a specialist in biology and environmental science in Instructional Materials Development. Dr. Morse has published extensively on molluscan biology (over 50 papers and 34 abstracts) and more recently in science education. She currently is a consultant for NSF, serving in the Division of Undergraduate Education as a site visitor to NSF-funded programs. Dr. Morse is a past president of Sigma Xi, the Scientific Research Society and the American Society of Zoologists (now the Society for Integrative and Comparative Biology). She is chair of the education committee of the American Institute of Biological Sciences and vice-chair of the International Union of Biological Sciences' Commission for Biological Education.

Margaret Cozzens is the Vice Chancellor for Academic and Student Affairs at the University of Colorado at Denver. She received a PhD in mathematics from Rutgers University. For the 7 years prior to coming UC Denver, she was the director of the Division of Elementary, Secondary, and Informal Education at the National Science Foundation (NSF). She also served as professor and chair of mathematics at Northeastern University in Boston before coming to NSF. She is the author of five books and over 75 papers and articles in mathematics and K-20 education policy. Dr. Cozzens currently serves on the President's Task Force on Teacher Education (American Council of Education). She co-chairs the Third International Assessment in Mathematics and Science 8th grade repeat study (TIMSS-R) Technical Review Panel. She also serves on the Education Advisory Council of the Department of Energy's National Renewable Energy Laboratory and the Task Force on Leadership for the Academic Affairs Resource Center of American Association of State Colleges and Universities.

Arthur Eisenkraft is the past President of the National Science Teachers Association, Arlington, Virginia. He is also the science coordinator (6-12) and physics teacher in the Bedford Public Schools in Bedford, New York. He has taught high school physics in a variety of schools for 24 years. He is project director of Active Physics and has published more than 100 papers related to physics and physics education. He holds a U.S. patent for a laser vision testing system. Dr. Eisenkraft is a former member of the National Research Council's (NRC) Advisory Panel to the Center for Science, Mathematics and Engineering Education (CSMEE). He served on the NRC Committee on Learning Research and Educational Practice and the NRC Working Group on Science Content Standards. He received his PhD in science education from New York University, and he received the Presidential Award for Excellence in Science Teaching in 1986.

Danine Ezell is a science teacher and department chair at the new charter school, Preuss School, chartered under the San Diego City Schools and associated with the University of California at San Diego. The Preuss School is designed to serve students from low-income and noncollege educated families and to prepare them for competitive 4-year colleges. Dr. Ezell previously worked as a resource teacher in the Mathematics and Science Office of San Diego City Schools and was the magnet school coordinator and teacher at Bell Junior High, a computer, mathematics, science magnet school in San Diego. Dr. Ezell is a former member of the NRC's Advisory Panel to CSMEE. She served on the NRC Working Group on Science Content Standards and worked for many years with Project 2061. She co-chaired her school district's efforts to develop science standards. She received a PhD in zoology from the University of California, Berkeley, and some years later obtained a teaching credential and began teaching at the secondary level in 1985.

Emily Feistritzer is the president of the Center for Education Information. She has conducted several national and state studies of alternative certification programs. She is currently conducting a study on the effectiveness of such programs. She has a PhD in curriculum and instruction from Indiana University, a masters in the teaching of science from the College of William and Mary, and a BA in biology from Thomas More College.

Maria Alicia Lopez-Freeman is the executive director of the California Science Project, a University of California professional development network of science faculty, professional developers and teachers of K-16 science. For many years she taught chemistry and physics in large, urban, inner-city high schools, developed programs, and served as department chair. For the past 10 years she has been working in science professional development, science education research, and educational change. She has published articles in both chemistry and science education and is currently involved in developing case studies focused on the intersection of science and equity in urban schools and doing research on science teaching and learning. She was a member of the Glenn Commission for the Teaching of Mathematics and Science for the 21st Century, and she serves on the High School Science Advisory Committee of the California Commission on Teacher Credentialing, developing program standards for the licensure of high school teachers of science. She was involved with the development of the *National Science Education Standards* as a member of the Working Group on Science Teaching Standards. Ms. Lopez-Freeman received her bachelor's and master's degrees from Immaculate Heart College.

Myles Gordon has been the vice president for education at the American Museum of Natural History in New York City since 1995. In that capacity he has engaged in a program with the City University of New York to retrain and to prepare science, math, and engineering professionals to teach in the New York City schools. Between 1970 and 1995 he worked in a variety of positions at the Education Development Center (EDC) in Newton, MA, becoming senior vice president in 1993. At EDC he was responsible for all work in science, math, and technology and led projects in curriculum development, professional development of teachers, instructional use of technology, and systemic reform. He is an advisory board member of the Association of Science and Technology Centers' Teacher Education Leadership Institute Project, the Annenberg/CPB Guide to Mathematics and Science Reform, and the Collaboration for Equity: Women in Science (a collaborative effort of the American Association for the Advancement of Science and EDC, funded by NSF). He has served as a consultant to the Corporation for Public Broadcasting's Regional Training Project, the Boston Public Schools, and the Educational Technology

Center/Harvard Graduate School of Education. He holds a BA in politics from Brandeis University.

Vicki Jacobs is the associate director of the Harvard University Teacher Education Programs and director of that institution's Mid-Career Math and Science Program (MCMS). Begun in 1983, the MCMS program has provided an alternative route to mathematics and science teacher certification in Massachusetts to mid-career scientists and mathematicians (some of whom hold Ph.D.s). She has also served as co-director of the Massachusetts Academy for Teachers.

David A. Kennedy was the director of science education/ESEA Title II, Office of the Superintendent of Public Instruction for the state of Washington. He is a biologist with degrees from Western Oregon University and Oregon State University. Prior to joining the superintendent's staff, he taught science subjects at the elementary, junior and high school levels, in teacher education programs at the college level, and he was a school district science coordinator. His assignments as an instructional program specialist at the state education agency have included supervisor of environmental education, 1971-1977; supervisor of science programs, 1977-1984; and senior supervisor of science and mathematics, 1984–1992. His responsibilities from 1992-1998 were as an instructional program director for which he managed a work unit of grants managers and curriculum specialists representing all content areas. He directs science education grants programs and develops the instructional design program. He is past president of the National Council of State Science Supervisors.

Mary Long is currently coordinator of UTeach, the secondary science and mathematics teacher preparation program at the University of Texas, Austin. She was a member of Office of Scientific and Engineering Personnel (OSEP)'s Committee on Attracting Science and Mathematics Ph.D.s to Secondary School Teaching. Ms. Long taught science for 28 years in middle schools and high schools in several states. She has served as director of the Austin Independent School District Science Academy and as manager of the district's Science and Health Resource Center. Ms. Long received a MEd degree in science education from the University of Texas, Austin.

John A. Moore is a professor emeritus of biology at the University of California at Riverside. He began serving on NRC education committees in the 1950s and continues to do so to this day. He is currently a member of the Committee on Science Education, K-12, and the National Science Resources Center's Advisory Board. His recent service to the NRC includes membership on the Committee on Undergraduate Science Education and several committees related to projects to produce materials for the K-12 teachers on the evolution/creationism problem. Beginning in the late 1950s, he worked with the Biological Sciences Curriculum Study, initially as the chair of the Committee of the Content of the Curriculum Study, then as supervisor for the high school textbook, "Biological Science: An Inquiry into Life." He also worked on two experimental and three commercial editions of the middle school project "Interactions of Man & the Biosphere" (1970-1979). He also initiated and supervised "Sciences as a Way of Knowing" —a university-level project that consisted of seven yearly symposia and publications (1983-1989) and edited the 17 volumes of the graduate-level series "Genes, Cells and Organisms: Great Books in Experimental Biology." Dr. Moore has served on the education committees of the NSF and of the AAAS (including Project 2061). Dr. Moore received AB, MA, and PhD degrees from Columbia University and has taught at Brooklyn College, Queens College, Barnard College, Columbia University, and the University of California at Riverside. He is a member of the National Academy of Sciences

N. Ronald Morris was chair of the OSEP Committee on Attracting Science and Mathematics Ph.D.s to Secondary School Teaching and is a professor in the Department of Pharmacology at the Robert Wood Johnson Medical School-University of Medicine and Dentistry of New Jersey (Rutgers). Dr. Morris is a cell biologist who studies nuclear migration and its regulation, using the fungus *Aspergillus* as a model organism. He has previously served as Associate Dean for Research at the Robert Wood Johnson Medical School—University of Medicine and Dentistry of New Jersey. He received an MD from Yale University School of Medicine and a BS from Yale College.

Kristina Peterson teaches chemistry and biology at the Lakeside School in Seattle, Washington. She received her PhD in analytical chemistry from the University of Washington in 1997. Before joining the Lakeside faculty, she served as a distance learning instructor, course designer, and as a science advisor to the Teacher Certification Program Wetlands Project at the University of Washington. She has been an active member of the Seattle chapter of the Association for Women in Science, serving as chair of its program committee, 1995-1996 and 1998-1999 and of its scholarship committee, 1994-1995. She was a member of OSEP's Committee on Attracting Science and Mathematics PhDs to Secondary School Teaching.

Eric Robinson, a professor at Ithaca College, received his PhD in mathematics from Binghamton University. His field of published mathematical research is topology. After receiving his doctorate, he taught at Bates College prior to joining the Department of Mathematics and Computer Science at Ithaca College 1979. He has served as interim associate dean of the School of Humanities and Science, and chaired the Department of Mathematics and Computer Science. In addition to his responsibilities at Ithaca, Dr. Robinson has also frequently taught preservice graduate-level content courses in mathematics in the master of arts in teaching program and Binghamton. He has done work in calculus reform, served as a program officer at the U.S. National Science Foundation in the Division of Elementary, Secondary, and Informal Education, and, since 1997, has been the director of COMPASS, the national secondary mathematics implementation center funded, in part, by NSF.

James H. Stith is the director of physics programs at the American Institute of Physics. He is past President of the National Society of Black Physicists and of the American Association of Physics Teachers. He is a fellow of the AAAS and the American Physical Society and a charter fellow of the National Society of Black Physicists. He was awarded the Distinguished Service Citation by the American Association of Physics Teachers, holds the New York Academy of Science's Archie L. Lacey Award for contributions to science education, the National Association for Equal Opportunity in Higher Education's (NAFEO) Distinguished Alumni Award, and was selected for Penn State University's Superior Teaching Award. He received his PhD from Pennsylvania State University and has served on the physics faculty of the U.S. Military Academy at West Point and Ohio State University.

Kimberly Tanner is a NSF postdoctoral fellow at the University of California at San Francisco in science education. She is working with the UCSF Science and Health Education Partnership to: (1) determine the factors that contribute to the success of teacher-scientist partnerships; (2) evaluate the impact of these partnerships on participating students, teachers, and scientists; and (3) utilize the results of this research to create materials to facilitate teacher-scientist partnerships that can be disseminated for use by universities and school districts nationwide. She received her PhD in neuroscience from the UCSF in 1997 and a BA in biochemistry/biology from Rice University in 1991. She was a member of OSEP's Committee on Attracting Science and Mathematics PhDs to Secondary School Teaching.